Heavy Metal Remediation

Scrivener Publishing
100 Cummings Center, Suite 541J
Beverly, MA 01915-6106

Publishers at Scrivener
Martin Scrivener (martin@scrivenerpublishing.com)
Phillip Carmical (pcarmical@scrivenerpublishing.com)

Heavy Metal Remediation

Edited by

Prangya Rath

Laxmi Kant Bhardwaj

and

Amit Kumar

Scrivener Publishing

WILEY

This edition first published 2027 by John Wiley & Sons, Inc., 111 River Street, Hoboken, NJ 07030, USA and Scrivener Publishing LLC, 100 Cummings Center, Suite 541J, Beverly, MA 01915, USA

For more information about Scrivener publications please visit www.scrivenerpublishing.com.

Wiley Global Headquarters
111 River Street, Hoboken, NJ 07030, USA

For details of our global editorial offices, customer services, and more information about Wiley products visit us at www.wiley.com.

The manufacturer's authorized representative according to the EU General Product Safety Regulation is Wiley-VCH GmbH, Boschstr. 12, 69469 Weinheim, Germany, e-mail: Product_Safety@wiley.com.

Library of Congress Cataloging-in-Publication Data

ISBN 978-1-394-40488-9

Cover image: Generated with AI using Adobe Firefly
Cover design by Russell Richardson

Set in size of 11pt and Minion Pro by Manila Typesetting Company, Makati, Philippines

Printed in the USA

10 9 8 7 6 5 4 3 2 1

Contents

Preface

The quick process of industrialization, agricultural development, and urbanization has played a critical role in human advancement. However, along with the advancements have come a hefty price for Mother Nature, as heavy metals now contaminate the environment. Heavy metals do not degrade like any other organic contaminants; rather, they continue to persist in our environment and eventually build up in the ecosystem of soils, waters, as well as our foods through the process of biomagnification. This is not simply a matter of environmental concern, but rather poses a direct and growing threat to humanity.

The idea of *Heavy Metal Remediation* came into being out of the pressing need to tackle this "Silent Threat." With the growing knowledge about the harmful impacts of heavy metals in the scientific community, it has been understood that conventional methods of cleaning up the environment have become insufficient due to their high disruptiveness. The requirement today is to look for ways to deal with these problems, which not only provide a solution but also ensure ecological sustainability, economic efficiency, and social justice.

The structure of this book is built with the goal of leading the reader to a step-by-step approach to understanding the issues involved and developing more sophisticated strategies for solving them. First of all, we need to provide some background on heavy metals and describe their pathways into the environment in detail. Since one can only manage that which can be measured, we start with some of the most recent advances in monitoring.

The key focus of the book is its discussion of sustainable remediation technologies that can actually be applied. Instead of focusing exclusively on the use of strong chemicals, we venture into the realms of sustainable science. Here, you will learn about chelation techniques that are applicable specifically to municipal solid waste compost, as well as the relatively new yet exciting science of bioremediation. You will see how nature can be used to counter pollution stresses as well as encourage plant growth.

Notably, technological development is impossible in isolation. In its analysis of the problem of Heavy Metal Contamination, Heavy Metal Remediation broadens its focus to include human and social aspects as well. We face the harsh challenges posed by developing countries, most importantly, the serious threats to human health and water resources due to dumpsite leachate. Through incorporating socio-economic factors into its discussion, the book recognizes that the efforts for remediation need to transcend barriers and seize opportunities in multiple dimensions.

In addition, the book also looks ahead to the future. As part of the final chapters, we explore the future outlook for global strategy, with multidisciplinary cases that explore the intriguing interface between nature and nanotechnology. The book is directed towards those environmentalists, engineers, agronomists, policymakers, and academics who are striving to restore our planet to health. It is our hope that by providing insights into methodologies and case studies here, both an essential source of academic information and inspiration for innovation can be achieved.

Editors
Dr. Prangya Rath
Dr. Laxmi Kant Bhardwaj
Dr. Amit Kumar

1

Overview of Heavy Metals

Shisak Sharma, Imotila T. Longchar, Raplang Steven Umdor and Dipak Sinha*

Department of Chemistry, Nagaland University, Lumami, Nagaland, India

Abstract

Heavy metal (HM) contamination has emerged as a critical global challenge, endangering public health and compromising environmental integrity due to its inherent toxicity and persistent nature in ecosystems. Stricter regulations on its discharge have focused the search for innovative, cost-effective, and sustainable remediation methods. Among the promising solutions, biosorbents derived from low-cost biowaste materials offer an environmentally friendly and economically viable approach to mitigating HM pollution. Activated carbon (C) synthesized from biowaste has garnered considerable attention for its effectiveness in wastewater treatment. Its production not only decreases the environmental burden of unused disposal but also enhances the affordability and efficiency of HM remediation. Thus, this chapter offers an inclusive appraisal of current development and use of biowaste-derived activated C for HM removal from contaminated water. The discussion also highlights the origins, toxicity, and environmental impact of HM pollutants, emphasizing the urgent need for effective mitigation. Additionally, it explores numerous remediation approaches, including chemical coagulation-flocculation, bioremediation, precipitation, phytoremediation, and adsorption. By integrating the utilization of biowaste with advanced wastewater treatment, this approach supports sustainable waste management practices and aligns with global initiatives aimed at conserving resources and protecting the environment.

***Keywords*:** Activated carbon, biomass, heavy metals,phytoremediation, pollution, remediation techniques

**Corresponding author*: dipaksinha@gmail.com; ORCID: https://orcid.org/0000-0002-4804-1640

Prangya Rath, Laxmi Kant Bhardwaj and Amit Kumar (eds.) Heavy Metal Remediation, (1–24)

1.1 Introduction

The reduction of accessible freshwater resources caused by increasing pollution levels contributes to worldwide water scarcity. Water contamination is a significant global issue that requires improved management at all levels. Water consumption grows at a geometric rate, while accessible water supplies have diminished due to inappropriate waste disposal. Consequently, global development presents a complex challenge regarding environmental conservation and protection. Pollution by HMs has been known as a major problem because of industrialization, urbanization, and globalization [1]. There are 35 as expected occurring metals, 23 of which are HMs because they have a specific density larger than 5.0 g/cm^3 and an atomic weight above 40.04 [2]. Alternatively, HMs can be described as metallic elements with a density at least five times greater than that of water [3]. HMs are a variety of elements with unique functions and chemical properties. They are classified as transition elements in the periodic table and can be categorized into both essential and non-essential metals. Essential metals serve an important part in the correct functioning of living organisms and are necessary in trace amounts. They are engaged in important biological processes like growth, metabolism, and organ development. However, when these metals are present in large concentrations, they can become poisonous and cause cellular damage. Non-essential metals, on the other hand, can impair living organisms by causing toxicity or carcinogenesis even at low doses. These metals are neither metabolized into intermediate chemicals nor degraded in the environment, making them persistent and possibly harmful [4].

Heavy metals (HMs) enter the ecosystem through a variety of pathways, including industrial, residential, agricultural, and medicinal activities, as well as natural phenomena such as volcanic eruptions and rock weathering [5]. These metals are toxic, non-biodegradable, and can remain in the environment for long intervals of time [6, 7]. Their presence presents significant risks to plants, animals, and humans, making them major environmental contaminants. HMs in the ecosystem follow a cyclical contamination cycle that includes industry, the food, soil, water, air, and human beings. They can enter the body through a variety of methods. Following absorption, which occurs mostly in the gastrointestinal tract, these water-soluble metals are transported to various organs *via* the circulatory system. HMs, even at minimal amounts, can disrupt key biological functions and cause major health concerns. Over the past few years, a substantial amount of work has been put into the development of efficient technologies for the removal of

HMs. These include ion exchange, chemical precipitation, electrochemical approaches, and adsorption [8–10]. A C-based adsorption technique has seen considerable application in the field of environmental remediation. Notably, activated C that is obtained from biomass is one of the most promising and cost-effective adsorbents for the removal of HMs [11].

A substantial amount of agricultural waste is produced annually on a global scale and is frequently disposed of without consideration, resulting in substantial resource waste. Effectively utilizing this trash can assist in reducing environmental constraints while also creating economic rewards. As a result, developing novel strategies for recycling agricultural waste is critical. Agricultural waste is the waste that is produced because of livestock reproduction, agricultural production, and sideline product processing [12]. This waste is primarily composed of remains in farmland and orchards, livestock dung, and residues in agricultural and marginal products and local waste [13]. Therefore, this study is to elucidate the environmental consequences and toxic effects of HMs. Additionally, the potential effects of specific HMs on living organisms, the properties of HMs, the sources, and the remediation methods of HMs contamination in the environment.

1.2 Heavy Metal (HM) Pollution

1.2.1 Sources of Heavy Metals (HMs)

Heavy metals (HMs) certainly exist in the environment and are vital to biological processes; however, they can become poisonous upon accumulation in living beings [14] (Figure 1.1). Among the HMs, mercury (Hg), arsenic (As), cadmium (Cd), nickel (Ni), chromium (Cr), lead (Pb), and copper (Cu) are some of the most prevalent environmental pollutants.

1.2.1.1 Mercury (Hg)

Mercury (Hg) is a highly toxic heavy element present in the biosphere. Human activities have rendered it a prevalent pollutant, resulting in its rising concentration in the atmosphere. Hg transforms into the very hazardous methylmercury (m-Hg) upon interaction with aquatic sediments. m-Hg enters the human body *via* the food chain *via* wildlife, fish, and seafood, which get infected following the absorption of hazardous bacteria. It enters the bloodstream following absorption into the human body and induces various neurological disorders [15].

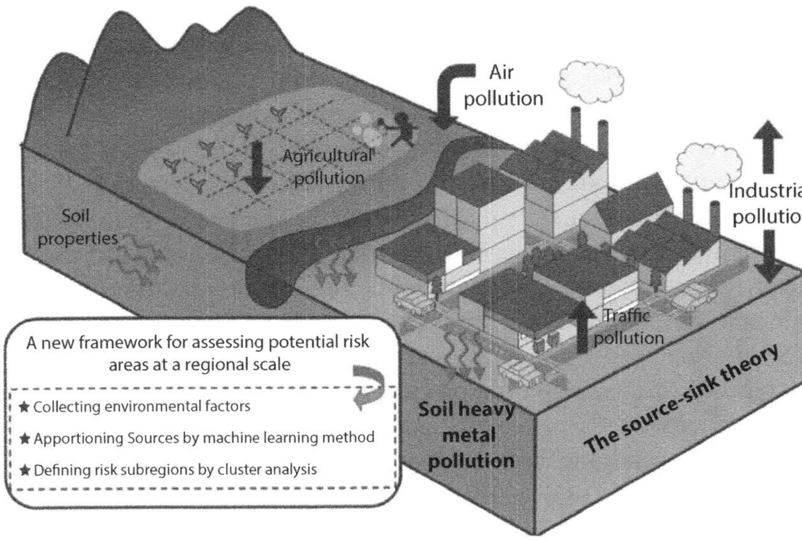

Figure 1.1 Schematic illustration of heavy metals (HMs) in the environment. Reproduced with permission from Reference [13]. Copyright Elsevier, 2020.

1.2.1.2 Cadmium (Cd)

Cadmium (Cd) is discharged into the atmosphere by natural or anthropogenic activities, resulting in varying exposure levels for animals and people. The contamination of Cd in aquatic ecosystems transpires *via* absorption, industrial effluents, and surface runoff into soil and sediments. Individuals can be poisoned by Cd through the intake of contaminated food, breath of polluted air, or ingestion of water with higher levels of the metal. Cd possesses no characteristics that benefit plant growth and metabolic functions [16].

1.2.1.3 Arsenic (As)

Arsenic (As) is predominantly emitted from diverse anthropogenic sources, including mining, fossil fuel burning, smelting, and chemotherapeutic agents for cancer treatment [17]. It is also emitted from paints and wood preservatives (Table 1.1).

1.2.2 Properties of Heavy Metals (HMs)

Metals are chemical elements distinguished by their great electrical and thermal conductivity, malleability, ductility, and reflectivity. Various

Table 1.1 Heavy metals (HM) sources and their detrimental effects on living things [18].

S. no.	Heavy metals (HMs)	Sources		Adverse impacts
		Natural	Anthropogenic	
1	Mercury (Hg)	Mineral cinnabar, volcanic eruptions, and fossil fuels, etc.	Mining, fossil fuel combustion, chlor-alkali processes, dental amalgams, etc.	Neurotoxicity, muscular weakness, blurred vision, speech and auditory impairments, premature mortality, etc.
2	Cadmium (Cd)	Volcanic activity, rock weathering, biogenic emissions, wildfires, etc.	Battery production, mining, smelting, refining of sulfide ores, etc.	Hypercalciuria, renal tubular injury, disruptions in calcium metabolism, etc.
3	Arsenic (As)	Volatilization, sedimentary rocks, pyrites, volcanic activity, geochemical processes, etc.	Wood preservatives containing arsenic (As), mines, landfills, pit heaps, smelter wastewater, etc.	Acute renal injury, impaired cardiac function, gastrointestinal complications, bone marrow damage, etc.

investigations have shown that metalloids tend to form covalent bonds, influencing their toxicological properties. HMs go through the human body by the intake of poisoned food and water, breath of polluted air, and skin exposure from residential, pharmaceutical, agricultural, and industrial sources. Metals are non-decomposable and incapable of decomposition [19]. Nevertheless, living organisms may detoxify metals *via* diverse processes, including sequestration inside proteins or deposition in insoluble granules. These purification strategies facilitate the elimination of metals through the organism's feces or their extended retention. Upon ingestion or breathing of HMs, they build up within our bodies over time, resulting in their designation as hazardous. This bioaccumulation results in physiological and biological consequences.

1.3 Heavy Metals (HMs) in Soil, Water, and Air

Environmental contamination may arise from metal corrosion, soil erosion, air deposition of metal ions, leaking of HMs, sediment re-suspension, and the loss of metals from water supplies to groundwater and soil. Natural events, including volcanic eruptions and weathering, have been documented to substantially contribute to HM contamination. Human activity influences the natural biological and geological restructuring of HMs through the contamination of soil, water, and air.

1.3.1 Air Pollution by Heavy Metals (HMs)

Air pollution is generally recognized as a dangerous form of pollution [20]. HMs from air pollution are a major worry in current times. Human activities show a substantial function in mobilizing HMs in the biosphere, contributing to their geochemical cycling. HM emissions from stationary and mobile sources in industrial locations exceed natural levels, affecting the environment, soil, and plants. HMs in the soil can produce airborne dust and particles, potentially impacting air conditioning. Persistent breathing of substantial quantities of HM particles may increase the body's burden of these metals and pose a health risk [21].

1.3.2 Soil Pollution by Heavy Metals (HMs)

Soils have exhibited significant contamination resulting from both the deposition of metals from vehicles and industries, as well as the mobilization of metals accumulated on road surfaces and near industrial areas. HM contamination in agricultural soils is escalating due to increased industrialization and urbanization in developing nations [22]. HMs constitute a primary group of inorganic pollutants, with a significant expanse of land contaminated by their presence due to the application of municipal compost or sludge, fertilizers, pesticides, and releases from municipal waste incineration, as well as exudates and remains from metalliferous mining and smelting industries. Soil dust and agricultural soil contamination by metals due to industrial activity, automobile emissions, and waste disposal sites are widely established. Alterations in soil qualities have been linked to environmental changes resulting from human activity [23].

1.3.3 Water Pollution by Heavy Metals (HMs)

Water pollution can be described in different ways. Typically, this indicates that one or more pollutants have accumulated in water to a degree that they negatively impact humans or animals [24]. All metals exhibit toxicity at elevated quantities, and their presence in water contributes to water pollution [25]. Elevated levels can be detrimental to the organism. HMs, including Hg, plutonium (Pt), and Pb, are poisonous substances with no recognized essential or advantageous effects on organisms. Their deposits in animal bodies over time can lead to severe health issues.

1.4 Mechanism of Heavy Metals (HMs) Toxicity

1.4.1 Mercury (Hg) Toxicity

Mercury (Hg) is well identified as a highly dangerous metal and is a predominant source of acute HM poisoning. m-Hg, a very detrimental neurotoxin, is recognized for inducing microtubule disintegration, mitochondrial impairment, fatty acid peroxidation, and the deposit of neurotoxic agents, including serotonin, glutamate, and aspartate [26]. The National Academy of Sciences and the Environmental Protection Agency have reported that between 8 and 10 % of American women have Hg levels high enough to cause neurological abnormalities in their children [27]. Animals subjected to hazardous concentrations of Hg have demonstrated considerable behavioral and neurological deficits. Rabbits exposed to Hg vapor concentrations of 28.8 mg/m^3 for 1 to 13 weeks demonstrated cellular deterioration, brain necrosis, and several nonspecific pathological changes [28]. Hg toxicity mostly affects the brain, yet it can also harm other organs, causing nervous system, renal, and muscle malfunction, etc.

1.4.2 Cadmium (Cd) Toxicity

The mechanism of Cd toxicity remains minimally understood; however, its cellular effects have been studied [29]. The concentration of Cd escalates 3000-fold upon binding to cysteine-rich proteins, such as metallothionein [30]. The cysteine-metallothionein compound in the liver prompts hepatotoxicity, then circulates to the kidneys, where it accumulates in renal tissue, resulting in nephrotoxicity. Cd can associate with ligands like cysteine, glutamate, histidine, and aspartate, potentially leading to iron (Fe) shortage [31]. Cd can replace zinc (Zn) in metallothionein because of its analogous oxidation state, impairing its role as a cellular free radical scavenger.

1.4.3 Arsenic (As) Toxicity

During biotransformation, deadly inorganic As compounds are methylated by organisms, including fungi, algae, bacteria, and humans, leading to the production of dimethylarsinic acid (DMA) and monomethylarsonic acid (MMA) [32]. In this biotransformation procedure, inorganic As (iAs) species are enzymatically transformed into methylation As compounds, which function as the definitive metabolites and serve as markers of chronic As exposure.

$$\text{iAs(V)} \rightarrow \text{iAs (III)} \rightarrow \text{MMA (III)} \rightarrow \text{DMA (V)}$$

Biomethylation functions as a detoxifying process, yielding methylated iAs molecules such as DMA(V) and MMA(V), which are eliminated *via* urine and serve as indicators for prolonged As exposure. MMA(III) is not eliminated and stays within the cell as an intermediate metabolite [33]. MMA(III), on the other hand, remains in the cell as an intermediate metabolite and is not eliminated. MMA (III), an intermediate compound, exhibits significantly more toxicity than other arsenicals and may be liable for As-induced cancer [34].

1.5 Effects of Heavy Metals (HMs) on Plants

Heavy metals (HMs) are naturally occurring in the earth's crust and are neither degraded nor destroyed. Life comprises both organic and inorganic elements; thus, the absence of metal ions renders life impossible. Numerous metals are important elements, such as Fe, cobalt (Co), Cu, selenium (Se), and Zn. They are present in plants at low concentrations and are essential for the maintenance of growth and metabolism. Nevertheless, when metals are present in concentrations above the needs of plants, they induce hazardous effects. The concentrations of some harmful HMs are presented in Table 1.2.

1.5.1 Effects of Mercury (Hg)

Mercury (Hg) is considered a unique metal since it exists in multiple forms in the environment, such as mercury-sulfide (HgS), Hg^{2+}, and m-Hg. The Hg^{2+} ion is the most abundant in agricultural soils. The ionic form of Hg is predominantly absorbed by soil and may also accumulate in aquatic vegetation. Hg^{2+} can cause physiological disturbances and impair metabolic

Table 1.2 Range of some hazardous heavy metals (HMs).

S. no.	Heavy metals (HMs)	Range in soils (ppm d. wt.)	Range in crops (ppm d. wt.)
1	Mercury (Hg)	0.01 to 0.2	0.005 to 0.02
2	Cadmium (Cd)	0.01 to 0.7	0.2 to 0.8
3	Arsenic (As)	10.0 to 40.0	0.01 to 5.0

processes in plants. It induces the excessive production of metabolites such as ascorbic acid, which can damage plant cells and disrupt the synthesis of chlorophyll and anthocyanin. The toxicity of Hg is contingent upon its chemical form, with the organic variant (m-Hg) exhibiting more toxicity than the ionic form (Hg^{2+}) [35]. Hg adversely affects seed germination and embryo development by creating S-Hg-S connections *via* its interaction with sulfhydryl groups (SHGs). The toxic effects result in diminished biomass, suppressed shoot growth, and alterations in plant shape. Hg exposure leads to a reduction in chlorophyll concentration due to compromised absorption of vital photosynthetic elements such as Fe and manganese (Mn), as well as damage to the photosynthetic apparatus [36].

1.5.2 Effects of Cadmium (Cd)

Cadmium (Cd) is a non-critical metal that adversely influences plant development and growth. It is known as a significant environmental contaminant because of its high water solubility and toxicological effects, even at minimal doses. Cd concentrations in uncontaminated soils generally range from 0.04 to 0.32 mM; however, contaminated soils may exhibit ranges from 0.32 to nearly 1.0 mM [37]. Cd toxicity is mainly characterized by symptoms including stunted development, chlorosis, and leaf rolling. It stimulates the production of oxygen (O_2) free radicals while diminishing the amounts of both non-enzymatic and enzymatic antioxidants. The roots are more likely to experience phytotoxic effects, such as reddish-brown leaf discoloration, root browning, and leaf chlorosis, because HMs tend to concentrate there more extensively [38]. Cd exposure influences plant growth by elevating the dry mass to fresh mass ratio in all plant organs. In elevated quantities, Cd interferes with multiple physiological processes, such as stomatal closure, compromised nutrient absorption, reduced enzyme activity, and disruptions in photosynthesis and respiration, thereby obstructing total plant growth [4].

1.5.3 Effects of Arsenic (As)

Arsenic (As) is a non-essential element for plants (or flora) and, owing to its pervasive global distribution, constitutes a considerable environmental hazard. It is extremely poisonous to both flora and other living species [39]. It gets into plants with vital nutrients, negatively impacting growth, crop production, and seed germination. The mobility of water is essential for the absorption of metalloids, metals, and mineral nutrients. Nonetheless, numerous HMs have been documented to impede water flow in various plant species [40]. The molecular similarity between As^{+5} and phosphorus (P^{+5}) disturbs phosphate metabolism [4]. It also affects the concentration of other metals such as potassium (K), calcium (Ca), Mn, and Zn. It is detrimental to plants by various mechanisms, including the reduction of Arsenate to Arsenite within plant cells, which disrupts aerobic phosphorylation. It induces stunted growth, impedes root extension, diminishes leaf quantity and fruit yield, accelerates leaf senescence, and may occasionally result in plant mortality [41].

1.6 Effects of Heavy Metals (HMs) on Human Health

Heavy metals (HMs) naturally exist in the ecosystem and can be advantageous to humans at low doses. They provide essential roles in numerous biological functions. For example, Fe facilitates hemoglobin synthesis, Cu participates in O_2 and electron transport, Co assists in cellular metabolism, Mn modulates enzymatic activity, Se plays a role in hormone and antioxidant production, and Ni promotes cellular growth and is designated as an essential metal. Nevertheless, when HMs are present in elevated concentrations, they become harmful to humans (Table 1.3).

1.6.1 Effects of Mercury (Hg)

Mercury (Hg) is regarded as the most hazardous non-critical metal for humans. It appears in the natural environment in three forms: organic Hg, inorganic Hg, and elemental Hg. Elemental Hg is predominantly released into the atmosphere as vapor. Its vapor primarily influences the central nervous system (CNS), resulting in cognitive, sensory, and motor disturbances, manifesting as sleeplessness, memory impairment, and muscle twitching. Prolonged interaction with elemental Hg results in diminished concentration, impaired vision, and unstable gait [42]. High levels of Hg exposure can cause serious brain damage and even death. Hg can traverse

Table 1.3 Permissible limits of some hazardous heavy metals (HMs) on human health.

S. no.	Heavy metals (HMs)	Max. acceptable concentration (World Health Organization (WHO)) (ppm)	Drinking water (Environmental Protection Agency (EPA)) (ppm/ppb*)
1	Mercury (Hg)	0.001	2.0*
2	Cadmium (Cd)	0.003	0.005
3	Arsenic (As)	0.01	0.01

the placental barrier, resulting in adverse effects on fetal development, including cerebral damage, blindness, cognitive disability, and speech impairments. High quantities of Hg exposure can generate respiratory diseases such as pneumonia and pulmonary edema, along with numerous clinical symptoms of pulmonary toxicity. Chronic low-level exposure is linked to dermatological symptoms, including skin rashes, desquamation of the hands and feet, and erythema, as well as neurobehavioral disturbances like depression and tremors, particularly in children [43].

1.6.2 Effects of Cadmium (Cd)

Cadmium (Cd) is regarded as a very hazardous HM with considerable health consequences. Both the United States National Toxicology Program (USNTP) and the International Agency for Research on Cancer (IARC) have classed it as a human carcinogen, highlighting its capacity to cause cancer with continuous or elevated exposure [29]. Cd interferes with essential biological functions by obstructing the synthesis of ribonucleic acid (RNA), deoxyribonucleic acid (DNA), and proteins, and facilitates the conversion of normal epithelial cells into malignant cells, therefore demonstrating carcinogenic characteristics [44]. Chronic exposure to Cd in humans is linked to lung cancer and considerable kidney damage. Cd also disrupts key trace elements like Zn, Fe, Ca, Mg, and Se, resulting in functional deficits and morphological changes in numerous organs, along with disturbances in secondary metabolic pathways. Clinically, Cd exposure may lead to symptoms such as pneumonia, tiredness, fever, and chest pain, and in severe instances, can be lethal [45]. Women exhibit heightened susceptibility to Cd poisoning owing to augmented intestinal absorption of dietary Cd, leading to elevated amounts in biological matrices, including urine, blood, and renal cortex [45].

1.6.3 Effects of Arsenic (As)

Arsenic (As) is extremely harmful to humans. Inorganic arsenic has been designated as a carcinogen. Low to moderate quantities of As exposure result in renal and hepatic failure, diabetes, and neurological problems. Skin is deemed more vulnerable to As, resulting in dermatitis; women are more predisposed to As-induced skin problems than men [46]. Skin lesions, including keratosis, melanosis, and pigmentation, are hallmark manifestations of As poisoning. The brain is also a target organ for As poisoning, that induces neurological problems. Neurons possessing elongated axons and sensory nerves are more significantly impacted than those with shorter axons and motor nerves. The ability of neurons to detoxify reactive oxygen species (ROSs) and produce glutathione diminishes, resulting in discomfort, numbness in the soles of the feet, and paraesthesia [47].

1.7 Remediation and Management Strategies

The environmental impacts of HMs are a major global issue owing to their considerable toxicity, persistence, and non-biodegradable characteristics. The removal of HMs from contaminated ecosystems has become a primary concern for both industrial stakeholders and scientific researchers. A diverse array of techniques and technologies has been established to remove HM concentrations, as shown in Figure 1.2.

All the approaches discussed above fall into one of three main categories: physical, biological, or chemical. The selection of treatment for HMs in a specific context is contingent upon various aspects, including the characteristics of the HMs, the permissible limits, and the concentrations of influent and effluent. The applicability of the various approaches is contingent upon removal efficiency, environmental impact, and economic viability [48].

1.7.1 Chemical Coagulation-Flocculation

Chemical coagulation-flocculation can be utilized to eliminate hazardous HMs and suspended particulates [49]. Chemical coagulants are introduced to the wastewater, causing the destabilization of colloidal fragments into aggregates. Flocculation is often used after the coagulation procedure

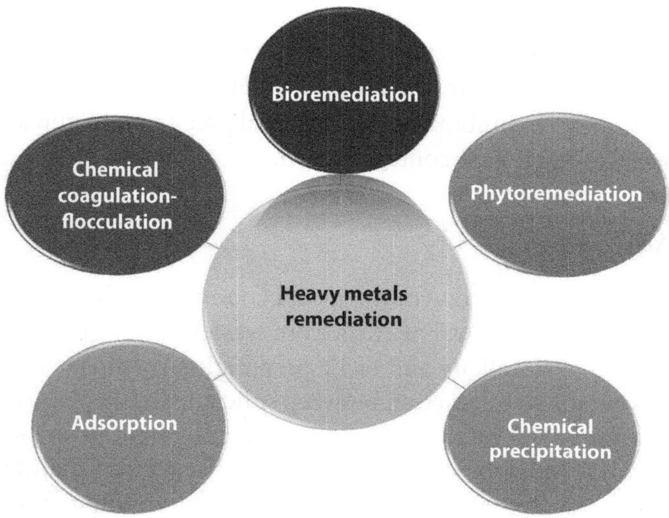

Figure 1.2 Treatment techniques for removal of heavy metals (HMs).

to aggregate destabilized fragments into bigger entities. Subsequently, sedimentation and filtering are used to extract the bigger fragments. A range of handling chemicals exists for wastewater management, including coagulants like ferric chloride ($FeCl_3$), aluminum sulfate ($Al_2(SO_4)_3$), and ferrous sulfate ($FeSO_4$), as well as flocculants such as polyelectrolytes, including polyacrylamide, polyferric sulfate, and polyaluminum chloride [50]. The efficacy of coagulation is influenced by factors like the kind of coagulant, coagulant strength, mixing circumstances, alkalinity, pH, and temperature. Although flocculation and coagulation are efficient remediation strategies for the elimination of HMs, it is frequently needed to augment this procedure with additional procedures to attain comprehensive removal of HMs. These approaches also result in the production of by-products that may function as secondary pollutants [51].

1.7.2 Chemical Precipitation

The chemical precipitation method involves the reaction of a precipitant with HM ions to form metal salts of poor solubility, which are subsequently separated from the aqueous solution using centrifugation or filtration, therefore facilitating the removal of ions from wastewater [52]. The elimination of HMs from aqueous solutions utilizing common chemical precipitation techniques involving lime, soda ash, and sodium sulfide (Na_2S) was compared. An original intensity of 100.0 mg/L resulted in a

99.9 % elimination of Zn and Cu from aqueous solutions using three precipitants, while Na_2S effectively removed 99.8 % of Pb. These precipitants have shown high efficacy in the removal of HMs [53]. The next research will concentrate on a combination of chemical precipitation and additional techniques to mitigate HM contamination.

1.7.3 Adsorption

Adsorption is considered one of the most economical, efficient, and selective methods for the removal of HMs from wastewater [54]. It is preferred for its operational simplicity and facile preparation methods. The adsorption process entails an interaction between solid and liquid phases, wherein HM ions (adsorbates) migrate from the aqueous phase to the surface of a solid material (adsorbent). The ions are subsequently maintained on the adsorbent *via* physisorption or chemisorption bonding mechanisms. Activated C is one of the most widely employed adsorbents for the elimination of hazardous HMs, due to its substantial robust affinity and surface area for metal ions [55]. However, its utilization is frequently hampered by obstacles like as high production costs and restricted regeneration capability [31]. As a result, there is an increasing interest in utilizing low-cost, sustainable biosorbents sourced from agricultural and industrial residue as alternative materials for treating HM-contaminated wastewater. In this context, agricultural byproducts or biomass materials, such as rice husk, clay materials, wheat-based waste, and fruit or vegetable peels, are used in the remediation of environments contaminated with HMs (Table 1.4).

1.7.4 Bioremediation

The investigation of an alternate approach, such as bioremediation, is necessary due to the drawbacks of various remediation strategies, including their high cost, formation of contaminated leftovers, non-specificity for metal-binding capabilities, and great reagent requirements [63]. This method is a novel approach to the elimination of poisonous metals from aquatic environments, utilizing biological systems such as microorganisms, animals, and vegetation. It has demonstrated efficacy in the extraction of HMs from contaminated soils and wastewater, ensuring sustainability in the restoration of polluted systems and restoring the natural condition of the affected ecosystem, yielding enduring ecological advantages at minimal expense. Consequently, numerous writers have utilized phytoremediation and bioremediation to remove HMs from water in recent years. Living creatures are recognized for their ability to detoxify hazardous substances in the environment [64].

Table 1.4 Reported adsorption efficiency (%) of several heavy metals (HMs) using adsorbents produced from biomass.

S. no.	Adsorbents	Adsorbate	Percentage (%)	References
1	Rice husk	As^{5+}	85.0 %	[56]
2	Wheat straw	Cd^{2+}	>80.0 %	[57]
3	Coffee husk	Cr^{6+}	97.4 %	[58]
4	Polish peat	Ni^{+2}	66.0 %	[59]
5	Groundnut shell	Cu^{+2}	77.8 %	[60]
6	Corn cob	Hg (II)	97.2 %	[61]
7	*Tectona grandis* sawdust	As (III)	75.0 %	[62]

1.7.5 Phytoremediation

Phytoremediation, often known as vegetative or green remediation, is an *in-situ* remediation technology that uses plants and allied microbes to immobilize, degrade, or convert environmental contaminants into non-toxic forms [53]. Although *ex-situ* testing is possible, it is not as effective as *in-situ* testing. Autotrophs may break down organic molecules and extract and stabilize inorganic metals in their aboveground biomass, making it relevant to both organic and inorganic pollutants [65]. Several plants demonstrate a superior potential for metal accumulation, making them suitable for remediating soils contaminated with harmful HMs. Phytoremediation is regarded as a cost-effective and environmentally sustainable method for remediating metal contaminants from contaminated sites, and it has garnered precedence over other methods in recent decades [66]. This approach is noteworthy as it provides site rehabilitation and preserves the biological and physical properties of soils, while also enabling the potential re-extraction of metals from plant biomass, a process known as phytomining of metals [67]. Phytoremediation employs three distinct methods for the remediation of HM-contaminated soils: phytoextraction, phytostabilization, and phytovolatilization (Figure 1.3).

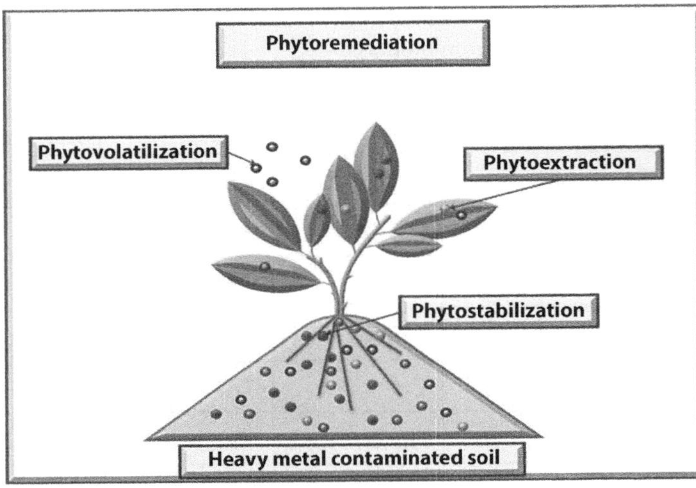

Figure 1.3 The mechanisms of heavy metal (HM) phytoremediation. Reproduced with permission from Reference [68].

1.8 Recommendations

Regulatory authorities must formulate and implement policies to restrict emissions and pollution from industrial sources. HM contamination can be decreased by enforcing stronger regulations on mining operations and industrial emissions. Monitoring should be a continuous process rather than a one-time effort. Promoting safe procedures to reduce exposure and educating the public about the dangers of HM are essential. Community engagement is essential in tackling this issue at both the individual and public levels. Monitoring at-risk populations and habitats on a regular basis is vital for early detection and mitigation of contamination. Personal protective equipment (PPE) should be made available to workers in occupations where exposure to dangers is possible. Regular health checkups should be conducted by authorized medical professionals.

1.9 Conclusion and Future Prospects

This chapter concluded that HM pollution is a major problem in the environment that endangers the environment and human health. The current study has elucidated the sources of HM pollution, the characteristics of a few HMs, and the health consequences associated with these harmful substances. This study consolidates the harmful impacts of HMs on both plants and human health, serving as a valuable resource for researchers studying HMs. It has also covered several techniques, including physical, chemical, and biological approaches, for the removal and control of HMs contamination. Despite numerous technologies for HM removal being proposed in recent years, many of these approaches are unsuitable for industrial application due to inefficiency, excessive energy intake, or inadequate selectivity. Activated C produced from biomass materials exhibits promising potential owing to its extensive definite surface area, intricate pore organization, great stability, and low cost. Phytoremediation, known as green or vegetative remediation, has been thoroughly examined and included in this study as a sustainable method utilizing plants and their associated microbial communities to immobilize, degrade, or convert HMs from the environment into non-toxic compounds.

Research on novel approaches to regulating and lessening the impact of HM pollution is critical because this detrimental phenomenon persists and endangers both human beings and ecosystems. A promising research domain is the application of nanotechnology (NT) to create novel and enhanced techniques for the detection, extraction, and remediation of HM. Furthermore, further research is necessary to elucidate the long-term effects of HM pollution on ecosystems and human health. An additional significant study domain is the advancement of sustainable agriculture techniques that minimize the utilization of fertilizers and herbicides containing HM. This can be accomplished through organic farming techniques that utilize natural processes to enhance soil condition and fertility. It is necessary to acknowledge the necessity of international collaboration and cooperation in tackling the problem of HM contamination.

List of Abbreviations

$Al_2(SO_4)_3$	Aluminum sulfate
As	Arsenic
C	Carbon
Ca	Calcium
Cd	Cadmium
CNS	Central nervous system
Co	Cobalt
Cr	Chromium
Cu	Copper
DMA	Dimethylarsinic acid
DNA	Deoxyribonucleic acid

EPA	Environmental Protection Agency
Fe	Iron
$FeCl_3$	Ferric chloride
$FeSO_4$	Ferrous sulfate
Hg	Mercury
HgS	Mercury-sulfide
HM	Heavy metal
IARC	International Agency for Research on Cancer
iAs	Inorganic As
K	Potassium
m-Hg	methylmercury
MMA	Monomethylarsonic acid
Mn	Manganese
Na_2S	Sodium sulfide
Ni	Nickel
NT	Nanotechnology
O_2	Oxygen
P	Phosphorus
Pb	Lead
PPE	Personal protective equipment
Pt	plutonium
RNA	Ribonucleic acid
ROSs	Reactive oxygen species
Se	Selenium
SHGs	Sulfhydryl groups
USNTP	United States National Toxicology Program
WHO	World Health Organization
Zn	Zinc

AI disclosure statement: The author acknowledges the use of AI-based tools for minor language editing or grammatical correction. However, all intellectual contributions, analyses, and conclusions presented in this chapter are entirely the author's own.

References

1. Liang, N., Zhang, C., Zhang, Y., Jiang, N., Yang, J., Geng, N., Zhang, R., Zhang, H., Chen, J., Pollution levels, distribution characteristics and risk assessment of heavy metals (HMs) in road dust in major cities of China. *J. Hazard. Mater.*, 494, 138733, 2025.

2. Gupta, V.K. and Ali, I., *Environmental Water: Advances in Treatment, Remediation and Recycling*, Elsevier, The Netherlands, 2012.
3. Hawkes, S.J., What is a "heavy metal"? *J. Chem. Educ.*, 74, 1374, 1997.
4. Bharti, R. and Sharma, R., Effect of heavy metals: An overview. *Mater. Today Proc.*, 51, 880–885, 2022.
5. Jayakumar, M., Surendran, U., Raja, P., Kumar, A., Senapathi, V., A review of heavy metals accumulation pathways, sources and management in soils. *Arab. J. Geosci.*, 14, 2156, 2021.
6. Alam, S., Bhardwaj, L.K., Mallick, R., Rai, S., Estimation of heavy metals and fluoride ion in vegetables grown nearby the stretch of River Yamuna, Delhi (NCR), India. *Indian J. Environ. Prot.*, 43, 64–73, 2023.
7. Bhardwaj, L.K., Sharma, S., Jindal, T., Estimation of physico-chemical and heavy metals in the lakes of Grovnes & Broknes Peninsula, Larsemann Hill, East Antarctica. *Chem. Afr.*, 6, 2677–2694, 2023.
8. Yaashikaa, P.R., Kumar, P.S., Jeevanantham, S., Saravanan, R., A review on bioremediation approach for heavy metal detoxification and accumulation in plants. *Environ. Pollut.*, 301, 119035, 2022.
9. Zeng, W., Guo, W., Li, B., Wei, Z., Dionysiou, D.D., Xiao, R., Kinetics and mechanistic aspects of removal of heavy metal through gas-liquid sulfide precipitation: a computational and experimental study. *J. Hazard. Mater.*, 408, 124868, 2021.
10. Wang, Z., Tan, Z., Li, H., Yuan, S., Zhang, Y., Dong, Y., Direct current electrochemical method for removal and recovery of heavy metals from water using straw biochar electrode. *J. Cleaner Prod.*, 339, 130746, 2022.
11. Wang, B., Lan, J., Bo, C., Gong, B., Ou, J., Adsorption of heavy metal onto biomass-derived activated carbon. *RSC Adv.*, 13, 4275–4302, 2023.
12. Poggi-Varaldo, H.M., Agricultural wastes. *Water Environ. Res.*, 71, 737–785, 1999.
13. Jia, X., Fu, T., Hu, B., Shi, Z., Zhou, L., Zhu, Y., Identification of the potential risk areas for soil heavy metal pollution based on the source-sink theory. *J. Hazard. Mater.*, 393, 122424, 2020.
14. Sardar, K., Ali, S., Hameed, S., Afzal, S., Fatima, S., Shakoor, M.B., Bharwana, S.A., Tauqeer, H.M., Heavy metals contamination and what are the impacts on living organisms. *Greener J. Environ. Manag. Public Saf.*, 2, 172–179, 2013.
15. Gworek, B., Dmuchowski, W., Baczewska-Dąbrowska, A.H., Mercury in the terrestrial environment: a review. *Environ. Sci. Eur.*, 32, 2020.
16. Hayat, M.T., Nauman, M., Nazir, N., Ali, S., Bangash, N., Environmental hazards of cadmium: past, present, and future. Cadmium Toxicol. Toler. Plants Physiol. Remediat., 163, 2019.
17. Bhardwaj, L.K., Rath, P., Upadhyay, D., Singh, V., Arsenic Removal from the Environment Using Biological Methods, in: *Emerging and Innovative Arsenic Removal Technologies for a Sustainable Future*, pp. 129–142, Springer Nature Switzerland, 2025.

18. Kaushik, P., Khandelwal, R., Rawat, N., Sharma, M.K., Environmental hazards of heavy metal pollution and toxicity: A review. *Flora Fauna*, 28, 315–327, 2022.
19. Sharma, A., Kaur, M., Katnoria, J.K., Nagpal, A.K., Heavy metal pollution: A global pollutant of rising concern, in: *Toxicity and Waste Management Using Bioremediation*, pp. 1–26, IGI Global Scientific Publishing, 2016.
20. Hesketh, H.D., *Air pollution control: traditional hazardous pollutants*, CRC Press, Taylor & Francis Group, 2023.
21. Lee, N.W., Wang, H.Y., Du, C.L., Yuan, T.H., Chen, C.Y., Yu, C.J., Chan, C.C., Air-polluted environmental heavy metal exposure increases lung cancer incidence and mortality: A population-based longitudinal cohort study. *Sci. Total Environ.*, 810, 152186, 2022.
22. Xia, F., Zhao, Z., Niu, X., Wang, Z., Integrated pollution analysis, pollution area identification and source apportionment of heavy metal contamination in agricultural soil. *J. Hazard. Mater.*, 465, 133215, 2024.
23. Rahmonov, O., Środek, D., Pytel, S., Kupka, T., Makieieva, N., Mineral composition of *Chelidonium majus* L. and soils in urban areas. *Appl. Sci.*, 15, 4718, 2025.
24. Zamora-Ledezma, C., Negrete-Bolagay, D., Figueroa, F., Zamora-Ledezma, E., Ni, M., Alexis, F., Guerrero, V.H., Heavy metal water pollution: A fresh look about hazards, novel and conventional remediation methods. *Environ. Technol. Innovation*, 22, 101504, 2021.
25. Dehkordi, M.M., Nodeh, Z.P., Dehkordi, K.S., Khorjestan, R.R., Ghaffarzadeh, M., Soil, air, and water pollution from mining and industrial activities: Sources of pollution, environmental impacts, and prevention and control methods. *Results Eng.*, 23, 102729, 2024.
26. Reardon, A.M. and Bhat, H.K., Methylmercury neurotoxicity: Role of oxidative stress. *Toxicol. Environ. Chem.*, 89, 535–554, 2007.
27. Nowicka, B., Heavy metal–induced stress in eukaryotic algae-mechanisms of heavy metal toxicity and tolerance with particular emphasis on oxidative stress in exposed cells and the role of antioxidant response. *Environ. Sci. Pollut. Res.*, 29, 12, 16860–16911, 2022.
28. Ohiagu, F.O., Chikezie, P.C., Ahaneku, C.C., Chikezie, C.M., Human exposure to heavy metals: toxicity mechanisms and health implications. *Mater. Sci. Eng. Int. J.*, 6, 2, 78–87, 2022.
29. Qu, F. and Zheng, W., Cadmium exposure: mechanisms and pathways of toxicity and implications for human health. *Toxics*, 12, 6, 388, 2024.
30. Zakaria, Z., Zulkafflee, N.S., Mohd Redzuan, N.A., Selamat, J., Ismail, M.R., Praveena, S.M., Tóth, G., Abdull Razis, A.F., Understanding potential heavy metal contamination, absorption, translocation and accumulation in rice and human health risks. *Plants*, 10, 1070, 2021.
31. Yan, L.J. and Allen, D.C., Cadmium-induced kidney injury: oxidative damage as a unifying mechanism. *Biomolecules*, 11, 11, 1575, 2021.

32. Hamza, M., Alam, S., Rizwan, M., Naz, A., Health risks associated with arsenic contamination and its biotransformation mechanisms in environment: a review, in: *Hazardous Environmental Micro-pollutants, Health Impacts and Allied Treatment Technologies*, pp. 241–288, Springer Nature Switzerland AG, 2022.
33. Jaishankar, M., Tseten, T., Anbalagan, N., Mathew, B.B., Beeregowda, K.N., Toxicity, mechanism and health effects of some heavy metals. *Interdiscip. Toxicol.*, 7, 60, 2014.
34. Medina, S., Bolt, A.M., Zhou, X., Wan, G., Xu, H., Lauer, F.T., Burchiel, S.W., Arsenite and monomethylarsonous acid disrupt erythropoiesis through combined effects on differentiation and survival pathways in early erythroid progenitors. *Toxicol. Lett.*, 350, 111–120, 2021.
35. Chen, J., Hao, S., Bañuelos, G., Zhou, X., A quantitative review of the effects of Se application on the reduction of Hg concentration in plant: a meta-analysis. *Front. Plant Sci.*, 14, 1199721, 2023.
36. Singh, H., Kumar, D., Soni, V., Impact of mercury on photosynthetic performance of *Lemna minor*: A chlorophyll fluorescence analysis. *Sci. Rep.*, 13, 12181, 2023.
37. Zulfiqar, U., Ayub, A., Hussain, S., Waraich, E.A., El-Esawi, M.A., Ishfaq, M., Ahmad, M., Ali, N., Maqsood, M.F., Cadmium toxicity in plants: Recent progress on morpho-physiological effects and remediation strategies. *J. Soil Sci. Plant Nutr.*, 22, 1, 212–269, 2022.
38. Li, Y., Rahman, S.U., Qiu, Z., Shahzad, S.M., Nawaz, M.F., Huang, J., Naveed, S., Li, L., Wang, X., Cheng, H., Toxic effects of cadmium on the physiological and biochemical attributes of plants, and phytoremediation strategies: A review. *Environ. Pollut.*, 325, 121433, 2023.
39. Beniwal, R., Yadav, R., Ramakrishna, W., Multifarious effects of arsenic on plants and strategies for mitigation. *Agriculture*, 13, 2, 401, 2023.
40. Khatun, J., Intekhab, A., Dhak, D., Effect of uncontrolled fertilization and heavy metal toxicity associated with arsenic (As), lead (Pb) and cadmium (Cd), and possible remediation. *Toxicol.*, 477, 153274, 2022.
41. Mondal, S., Pramanik, K., Ghosh, S.K., Pal, P., Ghosh, P.K., Ghosh, A., Maiti, T.K., Molecular insight into arsenic uptake, transport, phytotoxicity, and defense responses in plants: a critical review. *Planta*, 255, 4, 87, 2022.
42. Wu, Y.S., Osman, A.I., Hosny, M., Elgarahy, A.M., Eltaweil, A.S., Rooney, D.W., Chen, Z., Rahim, N.S., Sekar, M., Gopinath, S.C., Mat Rani, N.N.I., The toxicity of mercury and its chemical compounds: molecular mechanisms and environmental and human health implications: a comprehensive review. *ACS Omega*, 9, 5100–5126, 2024.
43. Queiroz, M.I., Sales, M.V., Barros, E.D.S.S., D'Amato, F.O., Goncalves, C.M., Ursulino, J.S., Bueno, N.B., Marinho, C., Rocha, U., Aquino, T.M., Fonseca, E.J.S., Exposure to a contaminated environment and its relationship with human health: Mercury effect on loss of functionality and increased oxidative stress of blood cells. *J. Hazard. Mater.*, 492, 138088, 2025.

44. Bertin, G. and Averbeck, D., Cadmium: cellular effects, modifications of biomolecules, modulation of DNA repair and genotoxic consequences (a review). *Biochimie*, 88, 1549–1559, 2006.
45. Charkiewicz, A.E., Omeljaniuk, W.J., Nowak, K., Garley, M., Nikliński, J., Cadmium toxicity and health effects-a brief summary. *Molecules*, 28, 18, 6620, 2023.
46. Algieri, C., Pugliese, V., Coppola, G., Curcio, S., Calabro, V., Chakraborty, S., Arsenic removal from groundwater by membrane technology: Advantages, disadvantages, and effect on human health. *Groundwater Sustainable Dev.*, 19, 100815, 2022.
47. Monteiro De Oliveira, E.C., Caixeta, E.S., Santos, V.S.V., Pereira, B.B., Arsenic exposure from groundwater: environmental contamination, human health effects, and sustainable solutions. *J. Toxicol. Environ. Health B.*, 24, 3, 119–135, 2021.
48. Vareda, J.P., Valente, A.J.M., Durães, L., Assessment of heavy metal pollution from anthropogenic activities and remediation strategies: a review. *J. Environ. Manag.*, 246, 101–118, 2019.
49. Skotta, A., Jmiai, A., Elhayaoui, W., El-Asri, A., Tamimi, M., Assabbane, A., El Issami, S., Suspended matter and heavy metals (Cu and Zn) removal from water by coagulation/flocculation process using a new Bio-flocculant: Lepidium sativum. *J. Taiwan Inst. Chem. Eng.*, 145, 104792, 2023.
50. Tang, X., Zheng, H., Teng, H., Sun, Y., Guo, J., Xie, W., Yang, Q., Chen, W., Chemical coagulation process for the removal of heavy metals from water: a review. *Desalin. Water Treat.*, 57, 1733–1748, 2016.
51. Bratby, J., *Coagulation and Flocculation in Water and Wastewater Treatment*, IWA Publishing, 2016.
52. Singh, S., Kapoor, D., Khasnabis, S., Singh, J., Ramamurthy, P.C., Mechanism and kinetics of adsorption and removal of heavy metals from wastewater using nanomaterials. *Environ. Chem. Lett.*, 19, 2351–2381, 2021.
53. Chen, Q., Yao, Y., Li, X., Lu, J., Zhou, J., Huang, Z., Comparison of heavy metal removals from aqueous solutions by chemical precipitation and characteristics of precipitates. *J. Water Process Eng.*, 26, 289–300, 2018.
54. Rashid, R., Shafiq, I., Akhter, P., Iqbal, M.J., Hussain, M., A state-of-the-art review on wastewater treatment techniques: the effectiveness of adsorption method. *Environ. Sci. Pollut. Res.*, 28, 9050–9066, 2021.
55. Kaveeshwar, A.R., Kumar, P.S., Revellame, E.D., Gang, D.D., Zappi, M.E., Subramaniam, R., Adsorption properties and mechanism of barium (II) and strontium (II) removal from fracking wastewater using pecan shell based activated carbon. *J. Cleaner Prod.*, 193, 1–13, 2018.
56. Babazad, Z., Kaveh, F., Ebadi, M., Mehrabian, R.Z., Juibari, M.H., Efficient removal of lead and arsenic using macromolecule-carbonized rice husks. *Heliyon*, 7, 2021.

57. Mahmood-ul-Hassan, M., Suthar, V., Rafique, E., Ahmad, R., Yasin, M., Kinetics of cadmium, chromium, and lead sorption onto chemically modified sugarcane bagasse and wheat straw. *Environ. Monit. Assess.*, 187, 2015.
58. Berihun, D., Removal of chromium from industrial wastewater by adsorption using coffee husk. *J. Mater. Sci. Eng.*, 6, 6–11, 2017.
59. Charazińska, S., Lochyński, P., Burszta-Adamiak, E., Removal of heavy metal ions from acidic electrolyte for stainless steel electropolishing *via* adsorption using Polish peats. *J. Water Process Eng.*, 42, 2021.
60. Kiran, B.M., Srikantaswamy, S., Pallavi, H.V., Manoj, V., Tasneem, T., A study on utilization of groundnut shell as biosorbent for heavy metals removal. *J. Environ. Sci., Comput. Sci. Eng. Technol.*, 2, 173–186, 2013.
61. Liu, Z., Sun, Y., Xu, X., Meng, X., Qu, J., Wang, Z., Liu, C., Qu, B., Preparation, characterization and application of activated carbon from corn cob by KOH activation for removal of Hg (II) from aqueous solution. *Bioresour. Technol.*, 306, 123154, 2020.
62. Sahu, N., Singh, J., Koduru, J.R., Removal of arsenic from aqueous solution by novel iron and iron-zirconium modified activated carbon derived from chemical carbonization of *Tectona grandis* sawdust: Isotherm, kinetic, thermodynamic and breakthrough curve modelling, p. 111431, 2021.
63. Ayangbenro, A.S. and Babalola, O.O., A new strategy for heavy metal polluted environments: a review of microbial biosorbents. *Int. J. Environ. Res. Public Health*, 14, 94, 2017.
64. Dixit, R., Wasiullah, X., Malaviya, D., Pandiyan, K., Singh, U.B., Sahu, A., Shukla, R., Singh, B.P., Rai, J.P., Sharma, P.K., Lade, H., Bioremediation of heavy metals from soil and aquatic environment: an overview of principles and criteria of fundamental processes. *Sustainability*, 7, 2189–2212, 2015.
65. Algreen, M., Trapp, S., Rein, A., Phytoscreening and phytoextraction of heavy metals at Danish polluted sites using willow and poplar trees. *Environ. Sci. Pollut. Res.*, 21, 8992–9001, 2014.
66. Bhardwaj, L.K., Kumar, D., Kumar, A., Phytoremediation potential of Ocimum sanctum: A sustainable approach for remediation of heavy metals, in: *Phytoremediation potential of medicinal and aromatic plants*, pp. 46–57, Taylor & Francis Group, 2023.
67. Ali, H., Khan, E., Sajad, M.A., Phytoremediation of heavy metals-concepts and applications. *Chemosphere*, 91, 869–881, 2013.
68. Khalid, S., Shahid, M., Niazi, N.K., Murtaza, B., Bibi, I., Dumat, C., A comparison of technologies for remediation of heavy metal contaminated soils. *J. Geochem. Explor.*, 182, 247–268, 2017.

2

Tracing the Silent Threat: Unraveling Heavy Metal Contamination Pathways and Impacts across Environmental Matrices

Qudrat Ullah[1]*, Laxmi Kant Bhardwaj[2], Waqas Haider[1] and Tassaneem Yuttasa-Ard[1]

[1]Food Technology and Innovation Research Centre of Excellence, School of Agricultural Technology and Food Industry, Walailak University, Nakhon Si Thammarat, Thailand
[2]Amity Institute of Environmental Toxicology, Safety and Management, Amity University, Noida, Uttar Pradesh, India

Abstract
Heavy metals (HMs) such as lead (Pb), cadmium (Cd), and mercury (Hg) are ubiquitous contaminants that contaminate water, soil, and air due to industrial effluents, agricultural practices, and natural processes. In this chapter, all significant anthropogenic sources of Cd include mining and smelting (1.8 million tons of Pb and zinc (Zn) annually), sewage sludge irrigation, and coal combustion, and their contamination pathways and impacts in various environmental matrices are unraveled. HMs leach into groundwater and, as well, bind to soil and are released to air, bioaccumulate in organisms (Pb 1.84 mg/kg in gills), and inhibit soil fertility enzymes. HMs are toxic to humans and cause oxidative stress, deoxyribonucleic acid (DNA) damage, and cancer risks, with children disproportionately at risk. The crisis is exemplified through case studies from Shazand Plain, Iran, and Egypt's Nile. To abate this silent threat to ecosystems and health, the author advocates for an integrated approach that combines phytoremediation, biochar, and strict policies.

***Keywords*:** Bioaccumulation, contamination, environment, heavy metals, remediation, transport

**Corresponding author*: qudratullahmpur@gmail.com; ORCID: https://orcid.org/0000-0003-3881-0917

Prangya Rath, Laxmi Kant Bhardwaj and Amit Kumar (eds.) Heavy Metal Remediation, (25–44)

2.1 Introduction to Heavy Metal (HM) Contamination

Heavy metals (HMs) are persistently lurking in the quiet undercurrents of our environment. They are metallic elements with an atomic weight higher than 5.0 g/cm^3, which are environmentally recalcitrant [1]. They are non-biodegradable, which allows them to exist in a broad range of air, water, and soil matrices, where they are toxic at trace levels. For example, aquatic life is impaired by Cd, even as low as 0.5 μg/L in water, through disruption of enzymatic functions [2]. The existence of a plethora of pathological HMs results from both natural processes (e.g., geological weathering) and anthropogenic activities (e.g., industrial effluents) [3]. They're astonishingly widespread: in soil, where they bind to organic matter, making for less diverse microbial communities; in water, where they are downcycled into the sediments; in air, where they travel as particulate matter (PM) for unlucky populations to breathe. Some HMs, such as Cd, Pb, and Hg, remain in soil, water, and air, which represents a potential risk for ecosystems and human health [4, 5]. The persistence of HMs, coupled with bioaccumulation in organisms, amplifies their ecological footprint, making them a critical focus of environmental science [6].

The global reach of HM contamination is a narrative of silent devastation. In 2023, Wang [7] reported that Pb concentrations in Chinese agricultural soils average 45.0 mg/kg, exceeding safe limits by 20%, and threatening crop safety. In Bangladesh, groundwater levels reached 50.0 μg/L in 2024, affecting 30 million people and causing arsenical skin lesions [8, 9]. Cd inhibits soil nitrogen (N) fixation, reducing yields by up to 15%, and poses a public health crisis, with Pb linked to cognitive deficits in 5% of exposed children globally. In Africa, mining-related Hg emissions contaminated 200,000 km^2 of water bodies, decimating fish populations and livelihoods [10]. Solutions like phytoremediation and biochar application show promise, with studies demonstrating a 30% reduction in soil Cd bioavailability. Yet, challenges remain; inadequate monitoring and lax regulations exacerbate contamination. Being toxic by mechanisms of oxidative stress and DNA damage, HM remains an urgent call for science-driven intervention to secure human health and ecosystems.

2.2 Sources of Heavy Metals (HMs) in Water, Soil, and Air

The insidious spread of HMs across water, soil, and air weaves a tale of human ambition and natural processes colliding with dire consequences. Anthropogenic excess and geological inevitability both propel HMs into environmental matrices (from industrial heartlands to volcanic calderas), and Pb, Cd, and Hg are the case in point [11]. Due to their persistence and toxicity, their origins demand rigorous scrutiny, which threatens ecosystems and human health. Figure 2.1 shows how HMs enter soil, water, and air.

2.2.1 Anthropogenic Sources: Industrial and Agricultural Contributions

Contamination of HM occurs more due to human activities, since most activities in industry and agriculture contribute more than the

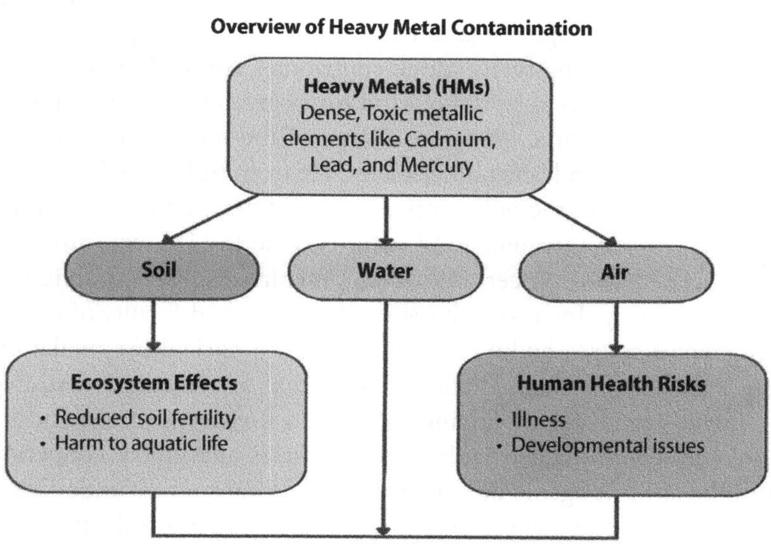

Figure 2.1 A simple illustration showing how heavy metals (HMs) enter soil, water, and air, leading to ecosystem damage and health risks.

natural sources [12]. Each year, the mining and smelting processes put 18 million tons of Pb and Zn into the soil and water of the planet [13]. Cd concentration levels of groundwater are found to increase to 15.0 μg/L even in the Hunan province of China, because of the smelting, up to a limit capable of exceeding even the World Health Organization (WHO) standard [14]. The Indian coal burning involves the release of 120 tons of Hg per annum, which is deposited on soils. Presently, the tanneries in Bangladesh release 21,000.0 m^3 of chromium (Cr) containing effluents per day, which contaminate rivers and agricultural lands [15]. Agricultural phosphate fertilizers increase the accumulation in Australian soils by 0.5 mg/kg Cd, while the sewage sludge irrigation in Europe is 1.2 million tons of nickel (Ni) and Pb, which disrupts the ecosystems [16]. HM loads are augmented by natural sources, such as geological weathering (10.0 to 1000.0 μg/L Cr in the Shazand Plain of Iran) or volcanic eruptions (5,000 tons Hg during the 2023 eruption of Eyjafjallajkull in Iceland), or even bushfires (2,000 tons Cd and Pb in Australia) [17, 18]. Nonetheless, anthropogenic emissions are prevalent, increasing the HM bioavailability, bioaccumulation in the food chains, and causing adverse health effects, i. e. neurological damage.

As Table 2.1 shows, anthropogenic HM emissions are on a staggering scale, and sewage sludge irrigation and mining dominate global contamination. China's mining sector alone contributes almost 2 million tons of Pb and Zn, adding up to be a thousand times greater than natural contributions such as volcanic emissions. Some of the regional disparities are stark; Cr from Bangladesh's tanneries and Hg from India's coal plants leave soils overloaded with the pollutants. Cr and Cd threaten aquatic biodiversity, and Pb in soil threatens food security. Due to HM exposure having been associated with cancer and developmental disorders, public health risks are escalating. Inconsistent regulations and the difficulty of monitoring non-point source pollution, especially from agriculture, make monitoring difficult. Phytoremediation and biochar as remediation strategies have shown promise in removing 25% of Cd toxicity in trials, and these technologies can be scaled with stricter policies and global cooperation. Finally, the anthropogenic dominance over natural sources underlines the need to nurture sustainable industrial practices and powerful environmental governance.

Table 2.1 Annual heavy metal (HM) release from anthropogenic sources.

S. no.	Type of sources	Heavy metals (HMs)	Estimated annual release (tons/year)	Environmental medium	References
1	Mining operations	Lead (Pb) and zinc (Zn)	1,800,000.00	Soil and water	[19]
2	Coal combustion	Mercury (Hg)	120.00	Air and soil	[20]
3	Tannery effluents	Chromium (Cr)	8,000.00	Water and soil	[21]
4	Phosphate fertilizers	Cadmium (Cd)	1,500.00	Soil	[22]
5	Sewage sludge irrigation	Nickel (Ni) and lead (Pb)	1,200,000.00		[16]
6	Smelting plants	Cadmium (Cd)	2,500.00	Water and soil	[23]
7	Battery recycling	Lead (Pb)	50,000.00		[24]
8	Pesticide application	Copper (Cu)	3,000.00		[25]

2.3 Pathways of Heavy Metal (HM) Distribution across Environmental Matrices

Heavy metals (HMs) have a relentless following through water, soil, and air in silence, following intricate physical and chemical processes. Transport mechanisms of HM (industrial outflows, wind-swept dust) shape environmental and human health risks and require a deep understanding of their transport.

2.3.1 Transport Mechanisms in Water and Soil

Heavy metals (HMs) like Cd and Pb are dynamic in water and soil. They enter food chains *via* soil, water, and air, accumulating in plants, animals, and humans [26]. Figure 2.2 traces their journey and growing concentration. HMs dissolve or form complexes in aquatic systems, and 10.0 to 20.0

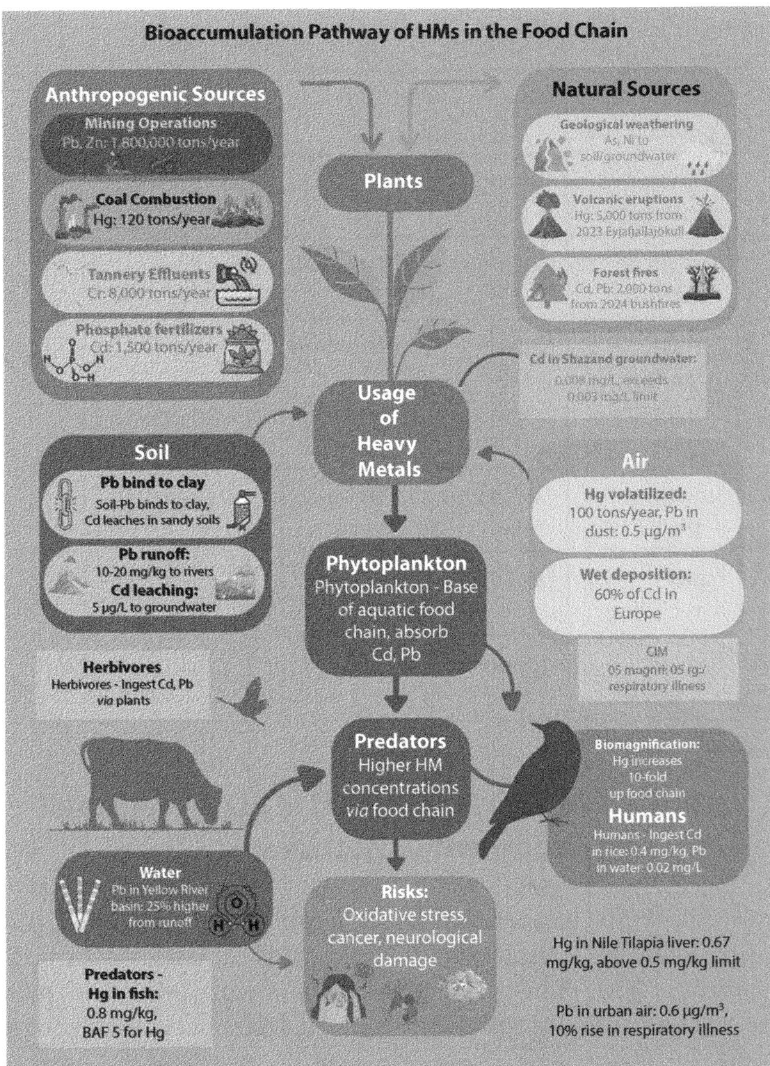

Figure 2.2 A diagram tracing the bioaccumulation of heavy metals (HMs) from environmental sources through plants and animals, intensifying risks at each step.

mg/kg of Pb is transported by runoff from contaminated soil into rivers annually. Cd is mobilized by leaching induced by rainfall; studies in India report values in groundwater near agricultural fields of 5.0 μg/L [27]. As an example, HM bioavailability is regulated by adsorption and desorption; the Pb is tightly bound to clay-rich soils and, therefore, Pb leaching is reduced, whereas Pb is readily leached into the sandy soils. HMs are also dispersed further *via* plant uptake, which in turn results in HMs taken up by rice roots at a rate of 0.3 mg/kg Cd and contributing to food chains [28]. The processes are affected by soil pH and organic matter; for example, acidic soils show an increase in Cd solubility by 15%, further magnifying the risks. Irrigation with HM-laden wastewater increases contamination, and, as found in this study, Pb levels in China's Yellow River basin are 25% greater due to industrial runoff [29]. The implication is that containing the spread of HM in terrestrial and aquatic ecosystems will be difficult.

2.3.2 Atmospheric Dispersion and Deposition

Heavy metals (HMs) are transported in the atmosphere as particles and vapors by wind and industrial emissions. In South Asia, 100 tons of Hg are volatilized from coal plants per year; Pb is carried vast distances by wind-borne dust [30, 31]. Dry and wet deposition remove HMs from the atmosphere into soils and waters, and wet deposition represents 60% of atmospheric Cd in Europe [32]. Deposition rates increase by 20% due to climate factors such as monsoon intensity in the tropics, and deposition rates are also elevated in urban areas near smelters. For Pb, particle-bound transport prevails, and urban air in Nigeria carries 0.5 μg/m^3, posing neurological risks [33]. The processes show that atmospheric HM dispersion is considerably far-reaching.

Equation 2.1: Heavy metal (HM) dispersion rate in air (mass balance model)

The Gaussian plume model quantifies HM dispersion in air.

Equation:

$$C(x,t) = \frac{Q}{\mu \cdot \sigma_y \cdot \sigma_z} \cdot \exp(-\frac{y^2}{2\sigma_y^2}) \cdot \exp(-\frac{(z-H)^2}{2\sigma_z^2})$$

Where

- $C(x, t)$: HM concentration (g/m^3) at distance (x) and time (t)
- Q: Emission rate (g/s)
- u: Wind speed (m/s)
- σ_y, σ_z: Horizontal and vertical dispersion coefficients
- H: Emission source height (m)

Melissano and CARPENTIERI [34] and Cañas Portilla *et al.* [35] cited this model, which predicts the spread of HM from sources such as smelters. For example, it has been used to estimate Cd concentrations 5 km from a factory in a 2023 European study, which reported 0.2 μg/m^3, guiding risk zoning. High-risk areas identified by the equation constrain policy; however, its effectiveness is limited by terrain complexity and variable meteorology, which change σ_y and σ_z by 10 to 20%. However, despite these constraints, these models are essential for transport modeling of HM and management of inhalation risk.

2.4 Impacts on Ecosystems: Soil Fertility, Aquatic Life, and Air Quality

Heavy metals (HMs) obstruct ecosystems, degrade soil vitality, aquatic life, and air quality. Insidiously, the presence of their activities, propelled by industrial and agricultural activities, acts as an interference in biological processes and biodiversity, and human health. HM impacts are occurring from barren fields to polluted waters and smog-filled skies, and their magnitude demands urgent scientific attention.

2.4.1 Effects on Soil Microbial Communities and Plant Growth

Heavy metals (HMs) not only destroy soil microbial communities and aquatic ecosystems, but also worsen the quality of the air [36]. Cd (10.0 mg/kg) decreases microbial biomass and inhibits urease and N cycling by 25%, whereas Pb minimizes *actinobacteria*, debilitating plant-microbe biotics by 15% [37]. Since (20.0 mg/kg) reduces wheat root elongation and chlorophyll by 30% and 18%, respectively, blocking photosynthesis. The Ni and Cr decrease the yields of maize in China by 12% [38]. The Cd bioavailability is reduced by 35% by having biochar, and a 20% increase in soil respiration is brought about by the interaction of biochar with Plant

Growth-Promoting Rhizobacteria (PGPR) [39]. *Brassica juncea*-based phytoremediation can extract 0.5 mg/kg Pb each year, but scaling up is problematic [40]. Hg, which exists in Indian carp at 0.8 mg/kg, biomagnifies 5-fold in aquatic environments, placing the nervous system at risk of damage [41]. Toxic levels of Cd (1.2 mg/kg) reduce the reproduction of oysters by 22% in mollusks. In the Shazand Plain of Iran, Cr (0.019 mg/L) deteriorates fish gills [42]. In Pakistan, Pb content in the atmosphere (0.6 $\mu g/m^3$) increases respiratory diseases by 10%, whereas acidities due to deposition of Hg (50.0 ng/m^2) affect the soils [43]. HM levels are lower in rural areas by 30%, but the urban zones encounter increased risks. Atmospheric HMs may be reduced by 25% by electrostatic precipitators and wetland restoration.

As shown in Table 2.2, HM levels are high in Shazand Plain concerning occupational and environmental protection standards, including soil threshold values (Pb, Cd, As, Cr, Ni, Hg, and iron (Fe)), as well as (Cd, Pb, As, Ni, and Hg) concerning groundwater standards. On the other hand, industrial and agricultural inputs are reflected in soil concentrations, especially Cr (35.28 mg/kg) and Ni (45.1 mg/kg), which may be risks to crops and microbial health. Although groundwater levels are low, leaching is indicated by low levels of Cd (0.008 mg/L, threatening the aquatic life. These excess stresses disrupt soil fertility, are bio-accumulated up food chains, and increase health risks, such as cancer, which calls for immediate remediation and rigid control on the pollution.

2.5 Human Health Effects: Bioaccumulation and Toxicity Risks

Modern societies are hit by silent invaders, HMs, and bioaccumulation of these pollutants, which drive a cascade of toxicological crises. Because HMs such as Pb, Cd, and As cross barriers to enter the body by ingestion, inhalation, and dermal contact, and cause oxidative stress and increased carcinogenic risk, adversely affecting vulnerable populations.

2.5.1 Pathways of Human Exposure: Ingestion, Inhalation, and Dermal Contact

Humans encounter HMs through multiple routes, each amplifying exposure risks. Ingestion dominates, particularly in regions like South Asia, where Cd in rice exceeds 0.4 mg/kg, surpassing WHO limits, and threatening

Table 2.2 Heavy metal (HM) concentrations in Shazand plain soil and groundwater.

S. no.	Heavy metal (HM)	Concentration in soil (mg/kg)	Concentration in groundwater (mg/L)	World health organization (WHO)/ Food and agriculture organization (FAO) permissible limit	References
1	Lead (Pb)	28.5	0.015	15.0 mg/kg; 0.01 mg/L	[44, 45]
2	Cadmium (Cd)	3.2	0.008	0.8 mg/kg; 0.003 mg/L	[45]
3	Arsenic (As)	12.7	0.012	10.0 mg/kg; 0.01 mg/L	[44]
4	Chromium (Cr)	35.28	0.019	20.0 mg/kg; 0.05 mg/L	
5	Nickel (Ni)	45.1	0.025	35.0 mg/kg; 0.02 mg/L	[45]
6	Zinc (Zn)	90.4	0.050	100.0 mg/kg; 5.0 mg/L	
7	Copper (Cu)	32.6	0.018	30.0 mg/kg; 2.0 mg/L	[44]
8	Mercury (Hg)	0.15	0.002	0.1 mg/kg; 0.001 mg/L	
9	Iron (Fe)	1500.0	0.300	1000.0 mg/kg; 0.3 mg/L	

20% of Bangladesh's population *via* dietary intake [46]. Contaminated drinking water, such as Pb levels of 0.02 mg/L in Pakistan's urban wells, further escalates risks, especially for children with higher ingestion rates [47]. Inhalation of HM-laden PM, with Pb concentrations reaching 1.8 $\mu g/m^3$ in Nigerian cities, endangers urban dwellers, particularly the elderly, by depositing metals in lung tissues [48]. Dermal contact, prevalent in Middle Eastern mining areas, exposes workers to Cr-rich soils (35.0 mg/kg), with absorption rates increasing under prolonged exposure. Pregnant women and children face heightened risks due to lower body weights and developing systems, amplifying HM uptake. These pathways get enhanced by industrial and agricultural runoff, emphasizing the need for source control along with a public awareness campaign.

2.5.2 Toxicological Impacts: Oxidative Stress and Carcinogenic Risks

Once inside the body, HMs unleash chaos. Bioaccumulation, where Cd in fish tissues reaches 0.5 mg/kg, and biomagnification, with Hg concentrations increasing 10-fold up aquatic food chains, concentrate toxins in organs like the liver and kidneys. Oxidative stress, marked by a 30% rise in malondialdehyde (MDA) and reactive oxygen species (ROS) in Pb-exposed populations, damages DNA and impairs neurological function [49]. As it induces lung cancer, a 2024 study reported elevated tumor markers in exposed Iranian communities. Cd's LD50 (oral, rats) is 225.0 mg/kg, and its No Observed Adverse Effect Level in humans (0.01 mg/kg/day) is frequently exceeded in contaminated regions, causing renal tubular damage. Mitigation strategies, including chelation therapy and phytoremediation, reduce HM bioavailability by 20 to 30%, but challenges like cost and scalability persist.

Equation 2.2: Bioaccumulation risk estimation (exposure dose model)
The Average Daily Dose (ADD) model quantifies HM exposure risks [50, 51].

Equation:

$$ADD = \frac{C \times IR \times EF \times ED}{BW \times AT}$$

Where

- *ADD*: Average Daily Dose (mg/kg/day)
- *C*: HM concentration (mg/kg or mg/L)
- *IR*: Ingestion rate (kg/day or L/day)
- *EF*: Exposure frequency (days/year)
- *ED*: Exposure duration (years)
- *BW*: Body weight (kg)
- *AT*: Averaging time (days)

In a 2021 Bangladesh study, *ADD* for as in drinking water ($C = 0.05$ mg/L, $IR = 2.0$ L/day, $EF = 350.0$ days/year, $ED = 30.0$ years, $BW = 60.0$ kg, $AT = 10{,}950$ days) was 0.00053 mg/kg/day, exceeding United States Environmental Protection Agency (USEPA's) reference dose (RfD) (0.0003 mg/kg/day) [52]. This yields a Hazard Quotient ($HQ = ADD/\text{RfD}$) of 1.77, indicating non-carcinogenic risk, and a Carcinogenic Risk ($CR = ADD \times$ Slope Factor) of 8×10^{-4}, signaling potential cancer risk. The model guides risk assessments but struggles with variable exposure patterns and children's higher *IR*. Stricter regulations and bioremediation are critical to curb these threats.

2.6 Case Studies: Shazand Plain, Iran, and Nile Tilapia in Egypt

Heavy metals (HMs) weave a toxic thread through ecosystems, with Shazand Plain, Iran, and Egypt's Nile Tilapia fisheries exemplifying regional crises. These case studies reveal how industrial and agricultural activities drive contamination, threatening soil, water, and food chains, with cascading effects on human health.

2.6.1 Shazand Plain: Groundwater and Soil Contamination Analysis

Nestled in Iran's Markazi Province, Shazand Plain is a hub of agriculture and industry, where oil refineries and phosphate fertilizers fuel economic growth but poison the land. Groundwater, the lifeblood of irrigation, carries Pb at 0.31 mg/L, exceeding WHO's 0.01 mg/L limit, while Cd reaches 0.008 mg/L, surpassing safe thresholds [44, 53]. Surface soils harbor Cd at 2.4 mg/kg, driven by vehicular emissions and fertilizer runoff, doubling FAO's 0.8 mg/kg guideline. Industrial discharges and heavy traffic, contributing 60% of Pb emissions, are the primary culprits. These HMs suppress soil microbial activity, reducing dehydrogenase enzyme levels by 22%, impairing nutrient cycling, and slashing wheat yields by 15%. Drinking contaminated water elevates blood Pb levels in locals, increasing the risk of renal damage by 18%. Recent risk assessments advocate biochar amendments, cutting Cd bioavailability by 30%, and stricter effluent controls to curb industrial pollution.

2.6.2 Nile Tilapia: Heavy Metal (HM) Accumulation in Aquatic Food Chains

Nile Tilapia (*Oreochromis niloticus*) is a dietary cornerstone in Egypt, but because of its role in the aquatic food chain, it acts as an HM magnet. The Damietta branch of the Nile is polluted from industrial effluents and agricultural runoff and its Tilapia Gills are tainted with Pb at the level of 1.84 mg/kg which exceeds the FAO limit of the Pb level in fish (0.3 mg/kg) and livers with Hg concentration of 0.67 mg/kg above WHO limit (0.5 mg/kg) on Hg content in fish [54, 55]. Muscle tissues, the primary edible portion, contain Cd at 0.41 mg/kg, posing risks to consumers. The tissue of Nile tilapia and the groundwater of Shazand have persistent Pb and Cd beyond safety levels. Comparisons of these critical concentrations are given in Table 2.3. Bioaccumulation factors (BAFs) for Pb reach 120 in gills, reflecting high uptake from contaminated water. This accumulation triggers oxidative stress, reducing tilapia reproduction by 25% and threatening fishery sustainability. For Egypt's fishing communities, consuming contaminated tilapia elevates cancer risk by 10^4, with children most vulnerable due to higher ingestion rates. Wetland restoration and phytoremediation are proposed solutions, reducing Hg levels by 20% in pilot studies.

Table 2.3 Comparative heavy metal (HM) concentrations in Shazand plain groundwater and Nile Tilapia tissues.

S. no.	Heavy metal (HM)	Sample type	Average concentration	References
1	Lead (Pb)	Groundwater (Shazand)	0.31 mg/L	[54, 56]
2	Cadmium (Cd)	Soil (Shazand)	2.4 mg/kg	[56]
3	Chromium (Cr)	Water (Shazand)	0.09 mg/L	[54]
4	Lead (Pb)	Gills (Tilapia)	1.84 mg/kg	[54, 56]
5	Mercury (Hg)	Liver (Tilapia)	0.67 mg/kg	[56]
6	Cadmium (Cd)	Muscle (Tilapia)	0.41 mg/kg	[54]

As shown in Table 2.3, Pb and Cd are the persistent HM threats in Shazand's groundwater and Egypt's tilapia tissues, consistently above the safe level. Industrial and agricultural pollutants feed excess HM into these ecosystems, with consequences keenly felt by humans living nearby through water and food, thereby endangering health. Bioremediation and policy enforcement provide solutions to break this toxic cycle.

2.7 The Need for Targeted Remediation Strategies

Industrial, agricultural, and geological compounds of HMs such as Pb, Cd, and As are distributed in water, air, and soil. Cr and Pb are contaminating groundwater in Shazand Plain, where they kill microbes in the soil. The Nile Tilapia of Egypt is contaminated with Hg; it poses a risk to the aquatic food chain. HMs slow down the activity of the soil enzymes, diminish aquatic life, and weaken the quality of the air, increasing the risks of cancer and neurological disorders, particularly in children. HM dispersion can be monitored by using real-time sensors and geospatial mapping. The solutions are through phytoremediation, biochar, and nanotechnology (NT) based filters. Sustainable agriculture is essential, as are strict industrial discharge limits. Unity and involvement of the community on a global scale guarantee clean water and food, protecting the ecosystems and human health.

2.8 Recommendations

Integrated approaches that employ phytoremediation technologies, biochar fertilization, and the implementation of NT-based filters should be used to fight against the problem of HM contamination to minimize the content of Pb, Cd, and Hg substances in the active soil layer, water, and air. The monitoring can be increased with real-time sensors and geospatial mapping in hotspots such as Shazand Plain and the Nile. Tougher control of industrial discharges and environmentally friendly farming, including limiting the use of phosphate fertilizers, is essential. Community participation, going hand in hand with global cooperation, will make clean resources accessible to all fairly. Policy incentives and biochar subsidies ($10/kg) can stimulate remediation at scale and solve the dual challenge of protecting ecology and human health.

2.9 Conclusion

Heavy metals (HMs) such as Pb, Cd, and As creep silently into soils, water, and airspace due to industrial and agricultural processes and natural processes. Pathways of and exposure to their dynamic pathways (leaching, deposition, and bioaccumulation) threaten soil microbes, aquatic life, and human health *via* cancer and neurological risks, and the authors reveal the pathways in this chapter. Hg in Nile tilapia is hazardous to food chains in Shazand plain, while Cr and Pb in the area exceed safe limits. Innovative approaches to remediate: real–time monitoring, geospatial mapping, and interdisciplinary solutions, including phytoremediation and biochar to extract or immobilize the HMs, are essential for effective HMs remediation. Industrial and agricultural rules must be rigorous to prevent contamination. Global cooperation and community engagement end the toxic cycle by offering equal access to clean resources. Effective, sustainable stewardship, which combines science and policy, is critical to providing a shield from this escalating ecosystem and human health crisis.

List of Abbreviations

ADD	Average Daily Dose
As	Arsenic
AT	Averaging time
BAFs	Bioaccumulation factors
BW	Body weight
Cd	Cadmium
CR	Carcinogenic Risk
Cr	Chromium
Cu	Copper
DNA	Deoxyribonucleic acid
ED	Exposure duration
EF	Exposure frequency
FAO	Food and Agriculture Organization
Fe	Iron
Hg	Mercury
HMs	Heavy metals
HQ	Hazard Quotient
IR	Ingestion rate
MDA	Malondialdehyde
N	Nitrogen

Ni	Nickel
NT	Nanotechnology
Pb	Lead
PGPRs	Plant-growth-promoting rhizobacteria
PM	Particulate matter
RfD	Reference dose
ROS	Reactive oxygen species
USEPA	United States Environmental Protection Agency
WHO	World Health Organization
Zn	Zinc

AI disclosure statement: The author confirms that no artificial intelligence (AI)-generated content or AI-assisted writing tools were used in the preparation of this chapter (including summaries and the abstract). All text, analysis, and conclusions are the result of the author's original work.

References

1. Naja, G.M. and Volesky, B., Toxicity and sources of Pb, Cd, Hg, Cr, As, and radionuclides in the environment, in: *Handbook of advanced industrial and hazardous wastes management,* pp. 855–903, CRC Press, Boca Raton, FL, USA, 2017.
2. Jiang, H., Li, R., Zhao, M., Peng, X., Sun, M., Liu, C., Liu, G., Xue, H., Toxic effects of combined exposure to cadmium and diclofenac on freshwater crayfish (Procambarus clarkii): Insights from antioxidant enzyme activity, histopathology, and gut microbiome. *Aquat. Toxicol.*, 268, 106844, 2024.
3. Masindi, V., Mkhonza, P., Tekere, M., Sources of heavy metals pollution, in: *Remediation of heavy metals,* pp. 419–454, Springer International Publishing, Cham, 2021.
4. Bhardwaj, L.K., Rath, P., Jain, H., Choudhury, M., Exploring the effects of e-waste on soil, water quality and human health. *Discov. Civ. Eng.*, 2, 1, 12, 2025.
5. Bhardwaj, L.K., Rath, P., Choudhury, M., Metzdorff, A., Jodder, P.K., Electronic wastes and contamination of water, in: *Electronic Waste*, pp. 48–57, CRC Press, Boca Raton, Florida, USA, 2025.
6. Edo, G.I., Samuel, P.O., Oloni, G.O., Ezekiel, G.O., Ikpekoro, V.O., Obasohan, P., Ongulu, J., Otunuya, C.F., Opiti, A.R., Ajakaye, R.S., Essaghah, A.E.A., Environmental persistence, bioaccumulation, and ecotoxicology of heavy metals. *Chem. Ecol.*, 40, 3, 322–349, 2024.
7. Wang, C.C., Zhang, Q.C., Yan, C.A., Tang, G.Y., Zhang, M.Y., Ma, L.Q., Gu, R.H., Xiang, P., Heavy metal (loid) s in agriculture soils, rice, and wheat

across China: Status assessment and spatiotemporal analysis. *Sci. Total Environ.*, 882, 163361, 2023.

8. Singh, P., Varshney, G., Kaur, R., Arsenic Contamination in Drinking Water and Health, in: *Arsenic Toxicity Remediation: Sustainable Nexus Approach*, pp. 125–142, Springer Nature Switzerland, Cham, 2024.
9. Tajwar, M., Rahman, M., Shreya, S.S., Sakib, N., Gazi, M.Y., Hasan, M., Islam, M., Alam, M.M.T., Zahid, A., Pollution evaluation and health risks assessment of naturally occurring toxic metals in shallow groundwater: A study in southwestern tidal delta of Bangladesh. *J. Trace Elem. Min.*, 10, 100197, 2024.
10. Ouma, K. O., Shane, A., Syampungani, S., Aquatic ecological risk of heavy-metal pollution associated with degraded mining landscapes of the Southern Africa River Basins: a review. *Minerals*, 12, 2, 225, 2022.
11. Younas, H., Nazir, A., Bareen, F.E., From sources to solutions: integrated approaches for Cd, Hg, and Pb remediation-a comprehensive review. *Plant Soil*, 510, 1, 1–47, 2025.
12. Hembrom, S., Singh, B., Gupta, S.K., Nema, A.K., A comprehensive evaluation of heavy metal contamination in foodstuff and associated human health risk: a global perspective, in: *Contemporary environmental issues and challenges in era of climate change*, pp. 33–63, Springer Singapore, Singapore, 2019.
13. Jagaba, A.H., Lawal, I.M., Birniwa, A.H., Affam, A.C., Usman, A.K., Soja, U.B., Saleh, D., Hussaini, A., Noor, A., Yaro, N.S.A., Sources of water contamination by heavy metals, in: *Membrane technologies for heavy metal removal from water*, pp. 3–27, CRC Press, Boca Raton, Florida, USA, 2024.
14. Pan, Y., Li, Y., Peng, H., Yang, Y., Zeng, M., Xie, Y., Lu, Y., Yuan, H., Relationship between groundwater cadmium and vicinity resident urine cadmium levels in the non-ferrous metal smelting area, China. *Front. Environ. Sci. Eng.*, 17, 5, 56, 2023.
15. Mizan, A., Assessment of contamination status and potential health risk of heavy metals in soil and vegetables of Savar Tannery adjacent area, 2020.
16. Nunes, N., Ragonezi, C., Gouveia, C.S., Pinheiro de Carvalho, M.Â., Review of sewage sludge as a soil amendment in relation to current international guidelines: A heavy metal perspective. *Sustainability*, 13, 4, 2317, 2021.
17. Kohzadi, S., Shahmoradi, B., Raushani, D., Nouri, A., INTERNATIONAL PERSPECTIVES: Evaluation and Risk Assessment of Heavy Metals in the Groundwater Resources of Saqqez, Iran. *J. Environ. Health*, 80, 6, E1–E9, 2018.
18. Pankratov, F., Mahura, A., Katz, O., Petäjä, T., Popov, V., Masloboev, V., Measurement report: Atmospheric mercury measurements at the Russian Arctic station Amderma and connection with eruptions of Icelandic volcanoes, Egusphere, pp. 1–18, 2025.
19. Sharifi, S.A., Zaeimdar, M., Jozi, S.A., Hejazi, R., Effects of soil, water and air pollution with heavy metal ions around lead and zinc mining and processing factories. *Water Air Soil Pollut.*, 234, 12, 760, 2023.

20. Cheng, B., Wang, Z., Yan, X., Yu, Y., Liu, L., Gao, Y., Zhang, H., Yang, X., Characteristics and pollution risks of Cu, Ni, Cd, Pb, Hg and As in farmland soil near coal mines. *Soil Environ. Health*, 1, 3, 100035, 2023.
21. Arti. and Mehra, R., Analysis of heavy metals and toxicity level in the tannery effluent and the environs. *Environ. Monit. Assess.*, 195, 5, 554, 2023.
22. Salim, S.A., Alalwany, A.A.M., Tawfeeq, G.H., The Influence of Industrial Residues of Phosphate Fertilizers Complexes on Content of Phosphor, Cadmium and Lead in Soil and Water. *IOP Conf. Ser.: Earth Environ. Sci.*, 1252, 1, 012054, 2023, IOP Publishing.
23. Adnan, M., Xiao, B., Xiao, P., Zhao, P., Li, R., Bibi, S., Research progress on heavy metals pollution in the soil of smelting sites in China. *Toxics*, 10, 5, 231, 2022.
24. Kumar, S., Rahman, M.A., Islam, M.R., Hashem, M.A., Rahman, M.M., Lead and other elements-based pollution in soil, crops and water near a lead-acid battery recycling factory in Bangladesh. *Chemosphere*, 290, 133288, 2022.
25. Alengebawy, A., Abdelkhalek, S.T., Qureshi, S.R., Wang, M.Q., Heavy metals and pesticides toxicity in agricultural soil and plants: Ecological risks and human health implications. *Toxics*, 9, 3, 42, 2021.
26. Angon, P.B., Islam, M.S., Das, A., Anjum, N., Poudel, A., Suchi, S.A., Sources, effects and present perspectives of heavy metals contamination: Soil, plants and human food chain. *Heliyon*, 10, 7, 2024.
27. Mahajan, M., Gupta, P. K., Singh, A., Vaish, B., Singh, P., Kothari, R., Singh, R.P., A comprehensive study on aquatic chemistry, health risk and remediation techniques of cadmium in groundwater. *Sci. Total Environ.*, 818, 151784, 2022.
28. Uddin, M.M., Zakeel, M.C.M., Zavahir, J.S., Marikar, F.M., Jahan, I., Heavy metal accumulation in rice and aquatic plants used as human food: A general review. *Toxics*, 9, 12, 360, 2021.
29. Li, J., Li, X., Wang, C., Liu, J.Z., Gao, Z.D., Li, K.M., Tuo, X.Y., Zang, F., Pollution characteristics and probabilistic risk assessment of heavy metal (loid) s in agricultural soils across the Yellow River Basin, China. *Ecol. Indic.*, 167, 112676, 2024.
30. Matei, E., Râpă, M., Mateş, I. M., Popescu, A. F., Bădiceanu, A., Balint, A. I., Covaliu-Mierlă, C.I. Heavy Metals in Particulate Matter—Trends and Impacts on Environment. *Molecules*, 30, 7, 1455, 2025.
31. Nguyen, L.S.P., Do, T.T.L., Vo, T.G. H., Le, Q.H., Hien, T.T., The Source and distribution of heavy metals in the atmosphere across southeast Asia, in: *Heavy Metal Remediation: Sustainable Nexus Approach*, pp. 1–26, Springer Nature Switzerland, Cham, 2024.
32. Liu, P., Wu, Q., Hu, W., Tian, K., Huang, B., Zhao, Y., Effects of atmospheric deposition on heavy metals accumulation in agricultural soils: Evidence from field monitoring and Pb isotope analysis. *Environ. Pollut.*, 330, 121740, 2023.

33. Ewona, I.O., Ekah, B.J., Akwagiobe, J.U., Udo, S.O., Rabiu, B., Evaluation of Hazard Quotient (Hq) of Toxic and Heavy Metal Concentration in Air Sediments Across Selected Cities in Nigeria, 2024.
34. Melissano, F. and Carpentieri, E., Real-time optimization of wind farm power through yaw steering control, 2021.
35. Cañas Portilla, A.I. and Castaño, A., World Health Organization., Human health effects of benzene, arsenic, cadmium, nickel, lead and mercury: Report of an expert consultation, World Health Organization, Regional Office for Europe, 2024.
36. Upadhyay, R., Heavy metals in our ecosystem, in: *Heavy Metals in Plants*, pp. 1–15, CRC Press, Boca Raton, 2022.
37. Chandwani, S., Kayasth, R., Naik, H., Amaresan, N., Current status and future prospect of managing lead (Pb) stress through microbes for sustainable agriculture. *Environ. Monit. Assess.*, 195, 4, 479, 2023.
38. Khan, S., Zahoor, M., Ullah, R., Khan, R. S., The uptake and mechanism of action of nanoparticles and doped nanoparticles on plant growth and metabolite enrichment. *Environ. Technol. Rev.*, 14, 1, 499–516, 2025.
39. Alshaal, T., Alharbi, K., Naif, E., Rashwan, E., Omara, A. E. D., Hafez, E.M., Strengthen sunflowers resilience to cadmium in saline-alkali soil by PGPR-augmented biochar. *Ecotoxicol. Environ. Saf.*, 280, 116555, 2024.
40. Rani, P., Rose, P.K., Kidwai, M.K., Meenakshi, Brassica Juncea L.: a potential crop for phytoremediation of various heavy metals, in: *Heavy metal toxicity: environmental concerns, remediation and opportunities*, pp. 285–311, Springer Nature Singapore, Singapore, 2023.
41. Le Vern, L., Biomonitoring of two marine mollusks, the Mediterranean mussel Mytilus galloprovincialis and the limpet Patella sp. in three North Corsican ports and inter-annual comparisons, 2022.
42. Taghavi, M., Shadboorestan, A., Kalankesh, L.R., Mohammadi-Bardbori, A., Ghaffari, H.R., Safa, O., Farshidfar, G., Omidi, M., Health risk assessment of heavy metal toxicity in the aquatic environment of the Persian Gulf. *Mar. Pollut. Bull.*, 202, 116360, 2024.
43. Rashid, S., Shah, I.A., Tulcan, R.X.S., Rashid, W., Sillanpaa, M., Contamination, exposure, and health risk assessment of Hg in Pakistan: A review. *Environ. Pollut.*, 301, 118995, 2022.
44. Sharafi, S. and Salehi, F., Comprehensive assessment of heavy metal (HMs) contamination and associated health risks in agricultural soils and groundwater proximal to industrial sites. *Sci. Rep.*, 15, 1, 7518, 2025.
45. Deng, X., Zou, H., Ren, B., Wang, J., Chen, L., Study on Pollution Characteristics, Sources, and Health Risks of Potentially Toxic Elements in Groundwater of Dongting Lake Basin, China. *Sustainability*, 17, 8, 2071-1050, 2025.
46. Afzaal, M., Hameeda, S., Liaqatb, I., Khanc, A.A.A., abdul Manand, H., Shahide, R., Altafe, M., corrected Proof, 2022.

47. Fida, M., Li, P., Wang, Y., Alam, S. K., Nsabimana, A., Water contamination and human health risks in Pakistan: a review. *Expo. Health*, 15, 3, 619–639, 2023.
48. Awos, A., Environmental assessment and impact of air, water and noise levels near a cement factory in Ewekoro, Ogun State, Nigeria, 2024.
49. Dey, K.K., Kamila, S., Das, T., Chattopadhyay, A., Chronic exposure to lead causes neurotoxicity by generating oxidative stress and inducing DNA damages in zebrafish brain: involvement of Nrf2-Keap1 regulation and DNA repair pathways. *Biol. Trace Elem. Res.*, 1–13, 2025.
50. Mirzaei, N., Kalteh, S., Zamani-Badi, H., Moradpour, H., Parmoozeh, Z., Baziar, M., Estimating human health risks associated with heavy metal exposure from bottled water using Monte Carlo simulation. *Heliyon*, 9, 10, 2023.
51. Emmanuel, U.C., Chukwudi, M.I., Monday, S.S., Anthony, A.I., Human health risk assessment of heavy metals in drinking water sources in three senatorial districts of Anambra State, Nigeria. *Toxicol. Rep.*, 9, 869–875, 2022.
52. Rahman, M.A., Kumar, S., Lamb, D., Rahman, M.M., Health risk assessment of arsenic, manganese, and iron from drinking water for high school children. *Water Air Soil Pollut.*, 232, 7, 269, 2021.
53. Sadeghi, S.H. and Hazbavi, Z., Land degradation in Iran, in: *Global degradation of soil and water resources: regional assessment and strategies*, pp. 287–314, Springer Nature Singapore, Singapore, 2022.
54. Shaalan, W.M., Hazardous Effects of Heavy Metal Pollution on Histological and Gene Expression Profiles of Nile Tilapia in the Eastern Delta, Egypt Aquatic Ecosystems, 2024.
55. Burch, E., Differential Levels of Nutrients and Heavy Metals in Tilapia Collected from Drains in Egypt, Master's thesis, The American University in Cairo, Egypt, 2022.
56. Suleman, R., Zahoor, M.A., Qarni, M.A., Saleh, I.A., Rao, W., Hussain, M., Ismail, T., Akhtar, S., Okla, M.K., Alaraidh, I.A., AbdElgayed, G., Assessment of heavy metals and microbial loads in Nile tilapia (Oreochromis niloticus) from different farms and rivers. *Sci. Rep.*, 15, 1, 5055, 2025.

3

Assessment and Monitoring Techniques for Heavy Metal Detection

Souravi Bardhan[1,2]* and Dipak Kr. Chanda[3]

[1]*Department of Environmental Science, Netaji Nagar College for Women, University of Calcutta, Kolkata, West Bengal, India*
[2]*Department of Physics, Jadavpur University, Kolkata, West Bengal, India*
[3]*Matrexia Advanced Technologies, North Subhas Pally, Universal Enclave, Dankuni, West Bengal, India*

Abstract
Quantifying heavy metals (HMs) in the environment, particularly those explicitly classified as hazardous, is critical for sound land-and water stewardship. Toxic HMs like lead (Pb), arsenic (As), mercury (Hg), chromium (Cr), and cadmium (Cd) are persistent and resistant to environmental degradation, and subsequently bioaccumulate, thus posing a serious threat to health, necessitating their monitoring in different matrices like air, water, soil, and biological samples. Conventional analytical platforms, including spectroscopy, spectrometry, X-ray fluorescence, etc., are benchmarks for trace quantitation and high specificity. The research continues to develop technology for portable devices that can rapidly detect and do so at a low cost using electrochemical sensors, nanomaterial (NM) based biosensors, and Laser-Induced Breakdown Spectroscopy (LIBS). Advanced spectroscopic methods like synchrotron X-ray absorption spectroscopy and spectrophotometric nanosensors provide higher selectivity and lower detection. However, they are currently faced with challenges of standardization, matrix interference, and scale-up. Thus, this chapter provides a summary of current and emerging techniques of HMs determination and their merits and limitations.

***Keywords*:** Biosensors, electrochemical sensors, heavy metals, optical sensors, real-time detection, spectroscopy

**Corresponding author*: bardhan666srv@gmail.com; ORCID: https://orcid.org/0000-0003-0944-2940

Prangya Rath, Laxmi Kant Bhardwaj and Amit Kumar (eds.) Heavy Metal Remediation, (45–82)

3.1 Introduction

Heavy metals (HMs) are a series of elements possessing metallic properties- great electrical conductivity (EC), malleability, and brightness, and comparatively high atomic weight (typically > 20) and great density (> 5.0 g/cm^3) [1]. HMs' toxicity is usually attributed to their lack of degradability, ability to bind high biological ligands such as thiol and imidazole groups of proteins, and biomagnifying and bioaccumulating properties towards food webs of the ecosystem [2, 3]. Anthropogenic activities, primarily mining, metallurgical processing, fossil fuel combustion, industry, urban runoff, and intensive agriculture, have perturbed the natural biogeochemical cycling of HMs [4]. Hg contamination in artisanal gold (Au) mining areas, Pb leaching from older water supply plants, As release from geogenic aquifers, and Cd soil deposition by phosphate fertilizers are just a few of the most significant environmental health concerns documented over the last decade. The United Nations Environment Program (UNEP) and World Health Organization (WHO) have also consistently put HMs in the list of the top ten chemicals of significant public health concern [5]. World monitoring reports currently cite the incidence of HM contamination globally [6].

Lead (Pb) and Cd concentrations have been reported to be high in over 70 % of the globe's industrialized river basins, among many of which WHO standard-conforming safe drinking water standards are grossly surpassed [7, 8]. Hg is transported within the air, enabling it to travel even to distant areas such as the Arctic, indicating that HM contamination is a transboundary problem [5]. In addition to this, bioaccumulation of these HMs in cereals, fish, and shellfish is a steady-state food pathway resulting in dietary exposure with neurotoxicity, nephrotoxicity, hepatotoxicity, immunotoxicity, and carcinogenesis [2]. Because of their persistence in the environment, mobility in certain physicochemical conditions, and severe toxicological impact of HMs, appropriate evaluation and monitoring are necessary to serve as guides for risk assessment, regulation, and cleanup [3].

Detection strategies have to be capable of handling a large number of analytical challenges like enormous concentration ranges (from ng/L in clean waters to mg/L in industrial wastewaters), challenging environmental matrices, and discriminating between total concentration and bioavailable or labile fractions [9]. Additionally, temporal and spatial fluctuations of contamination make it obligatory to incorporate high-resolution analytical instruments, *in situ* sensing instruments, and remote monitoring stations [9]. The pace of detection technology development is shifting from the venerable wet chemistry and atomic spectrometry through to biosensors, NM-based platforms, and hyperspectral remote sensing, which is a revolution towards field-portable, real-time, high-sensitivity [10]. These advances are required to bridge the gap between laboratory-scale analysis and persistent environmental monitoring to allow stakeholders to sense the new episodes of contamination while they still remain before they have caused irreparable ecological or public health disasters.

This chapter provides a critical review of existing and upcoming monitoring and assessment instruments for the analysis of HMs, from the more classical analysis methodologies, biosensing platforms, and nanotechnology (NT) based platforms, to new upcoming spectroscopic and remote sensing technologies. Integrating the science, technology, and sustainability aspects, this chapter attempts to highlight existing capabilities, operating limitations, and possible areas of future improvement in increasing the world's ability to detect and counteract HM pollution.

3.2 Sources and Ecological Impacts of Heavy Metals (HMs)

Heavy metals (HMs) are considered merely trace micronutrients needed by organisms. These toxic pollutants enter ecosystems from natural as well as anthropogenic sources, such as industrialization, urbanization, and intensive land use [2, 3]. Volcanic eruptions, weathering of metal-containing rocks, and forest fires are geogenic sources of HM load and are responsible for their flow to soil and water. Anthropogenic processes are the cause of most of this load. Metallurgical process and mining, electroplating, manufacture of batteries, tanning hides, dyeing cloth, petroleum refining, and incineration of wastes are some of the major industrial sources. Also, excessive application of phosphorus (P) fertilizers and pesticides in agricultural fields adds HMs into land systems that further migrate towards water bodies through surface runoff and percolation *via* groundwater [3, 8]. Various sources, effects, and simultaneous detection techniques have been illustrated in Table 3.1.

Aquatic species are subjected to adsorption of HMs onto sediment matrices and suspended phases, affecting aquatic food webs and microbial communities. They alter soil microbial diversity, inhibit enzymic processes, and inhibit the growth and productivity of plants through phytotoxicity. Bioaccumulation into the biota and biomagnification along

Table 3.1 Sources, effects, permissible limits, and detection techniques of various heavy metals (HMs).

S. no.	Heavy metals (HMs)	Major sources	Environmental and health effects	Associated diseases	Regulatory threshold limits	Common analysis techniques	References
1	Lead (Pb)	Battery recycling, leaded gasoline, paints, pipes, mining, and smelting	Neurotoxicity, kidney damage, anemia, and soil/water contamination	Plumbism (lead (Pb) poisoning), encephalopathy (in children), and peripheral neuropathy	*A: 0.01 mg/L #B: 400.0 ppm ^C: 0.05 mg/m^3	Inductively Coupled Plasma Mass Spectrometry (ICP-MS), Atomic Absorption Spectroscopy (AAS), and Anodic Stripping Voltammetry (ASV)	[7]
2	Cadmium (Cd)	Nickel-cadmium (Ni-Cd) batteries, phosphate fertilizers, metal plating, and coal combustion	Carcinogenic (lung cancer), kidney dysfunction, and bone demineralization	Itai-Itai disease (osteomalacia and kidney failure), and chronic obstructive pulmonary disease (COPD)	*A: 0.003 mg/L #B: 1.4 ppm ^C: 0.005 mg/m^3	Inductively Coupled Plasma Mass Spectrometry (ICP-MS), Graphite Furnace Atomic Absorption Spectrometry (GF-AAS), and X-ray Fluorescence (XRF)	[8]
3	Mercury (Hg)	Coal-fired power plants, gold (Au) mining, dental amalgams, and fish bioaccumulation (methylmercury)	Neurological damage, developmental defects (in fetuses), and kidney failure	Minamata disease, and acrodynia (pink disease)	*A: 0.006 mg/L #B: 2.0 ppm ^C: 0.025 mg/m^3	Inductively Coupled Plasma Mass Spectrometry (ICP-MS), Cold Vapor Atomic Absorption Spectrometry (CV-AAS), and Mercury (Hg) analyzer	[5]

(Continued)

Table 3.1 Sources, effects, permissible limits, and detection techniques of various heavy metals (HMs). (*Continued*)

S. no.	Heavy metals (HMs)	Major sources	Environmental and health effects	Associated diseases	Regulatory threshold limits	Common analysis techniques	References
4	Arsenic (As)	Pesticides, herbicides, mining, smelting, and natural groundwater contamination	Skin lesions, carcinogenic, and cardiovascular disease	Arsenicosis and blackfoot disease	*A: 0.01 mg/L #B: 0.4 ppm ^C: 0.01 mg/ m^3	Inductively Coupled Plasma Mass Spectrometry (ICP-MS), Hydride Generation-Atomic Absorption Spectrometry (HG-AAS), and Colorimetric methods	[11]
5	Chromium (Cr VI)	Leather tanning, electroplating, textile industries, and steel production	Carcinogenic, deoxyribonucleic acid (DNA) damage, and skin ulcers	Chromium (Cr) induced lung cancer, and nasal septum perforation	*A: 0.05 mg/L (Cr-VI) #B: 5.0 ppm ^C: 0.005 mg/ m^3	Inductively Coupled Plasma Mass Spectrometry (ICP-MS), Ultraviolet–visible (UV–Vis) Spectrophotometry, and Ion Chromatography	[12]

(*Continued*)

Table 3.1 Sources, effects, permissible limits, and detection techniques of various heavy metals (HMs). (*Continued*)

S. no.	Heavy metals (HMs)	Major sources	Environmental and health effects	Associated diseases	Regulatory threshold limits	Common analysis techniques	References
6	Nickel (Ni)	Stainless steel production, batteries, metal alloys, and industrial wastewater	Dermatitis, lung fibrosis, and carcinogenic potential	Nickel (Ni) Eczema (contact dermatitis), and lung fibrosis (chronic exposure)	*A: 0.07 mg/L #B: 160.0 ppm ^C: 1.0 mg/m^3	Inductively Coupled Plasma - Optical Emission Spectrometry (ICP-OES), Flame Atomic Absorption Spectrometry (AAS), and X-ray Fluorescence (XRF)	[13]
7	Copper (Cu)	Mining, electroplating, and agricultural fungicides - Plumbing corrosion	Liver destruction (Wilson's disease), gastrointestinal distress, and hemolytic anemia	Indian Childhood Cirrhosis and Wilson's Disease	*A: 2.0 mg/L #B: 310.0 ppm ^C: 0.1 mg/m^3	Inductively Coupled Plasma - Optical Emission Spectrometry (ICP-OES), Flame Atomic Absorption Spectrometry (AAS), and Colorimetric (Bathocuproine method)	[14]

(*Continued*)

Table 3.1 Sources, effects, permissible limits, and detection techniques of various heavy metals (HMs). (*Continued*)

S. no.	Heavy metals (HMs)	Major sources	Environmental and health effects	Associated diseases	Regulatory threshold limits	Common analysis techniques	References
8	Iron (Fe)	Steel industry, mining, rusting pipes, and natural weathering of rocks	Liver damage, oxidative stress, and soil/water discoloration	Hemochromatosis and Siderosis	*A: 0.3 mg/L #B: No specific limit (varies by region) ^C: 10.0 mg/m³ (as Fe_2O_3)	Inductively Coupled Plasma - Optical Emission Spectrometry (ICP-OES), Atomic Absorption Spectrometry (AAS), and Ultraviolet–visible (UV–Vis) Spectrophotometry	[12]
9	Zinc (Zn)	Galvanization, alloys, rubber manufacturing, and agricultural runoff	Growth retardation, nausea, vomiting, and immune suppression	Metal Fume Fever and Hypocupraemia	*A: 3.0 mg/L #B: 7,300.0 ppm ^C: 5.0 mg/m³	Inductively Coupled Plasma - Optical Emission Spectrometry (ICP-OES), Flame Atomic Absorption Spectrometry (AAS), and Ultraviolet–visible (UV–Vis) Spectrophotometry	[15]

*A = Drinking water (WHO); #B = Soil (residential) (United States Environmental Protection Agency (USEPA)); ^C = Air (Occupational Safety and Health Administration (OSHA)) food webs are highly detrimental to animal and human health, e.g., neurotoxicity, carcinogenicity, nephrotoxicity, and developmental aberrations [6, 8]. Furthermore, the HMs persist for centuries, and the application of natural attenuation processes is difficult to achieve; treatment involves rigorous monitoring and intervention. Their source and extent being certain is, therefore, necessitated to enable tailored detection and environment-friendly management practice.

3.3 Sampling and Pre-Treatment Procedures

For HMs, reliability and exactitude depend on sampling and pre-treatment of the highest grades. Environmental matrices like sediment, soil, water, and biological tissues are complex and heterogeneous, requiring extremely well-optimized techniques for analysis.

3.3.1 Sampling Strategies

The heterogeneity and complexity inherent in environmental systems require statistically representative and scientifically defensible sampling protocols. Samples should be sufficient to represent the matrix mean composition under investigation as well as the spatial contaminant distribution patterns.

3.3.1.1 Water Sampling

Prevention of analyte loss due to adsorption, precipitation, or volatilization is necessary in wastewater, surface water, and groundwater. Internationally accepted practices, such as the International Organization for Standardization (ISO) and the United States Environmental Protection Agency (USEPA), recommend using acid-washed Teflon or polyethylene (PE) vessels [16]. *In situ* acidification, usually using ultrapure nitric acid (HNO_3) to a $pH < 2$, quickly dissolves metal ions and inhibits microbiological activity. Samples should be stored at 4°C during transport. For point-source analysis, grab sampling at the source is appropriate, while composite sampling offers a better view of temporal variability [17].

3.3.1.2 Soil and Sediment Sampling

For solid matrices, grid-based systematic or random stratified sampling is used extensively to account for spatial heterogeneity. Stainless steel corers or augers allow sampling at particular depths to facilitate vertical profiling of contamination. The samples taken must be stored in inert packs and transported under dark, refrigerated conditions in order to prevent oxidative conversions and redox alterations due to microbial activity [16, 17].

3.3.1.3 *Biological Sampling*

Biological matrices such as plant tissues, animal organs, and microbial biofilms are becoming mainstream in biomonitoring programs. Sampling is by selective removal of the target tissue, soft rinsing in deionized water to remove particulate matter (PM) on the surface, and immediate freezing or lyophilization to maintain analyte integrity [16, 17]. All these processes must be done with ethical and biosafety protocols.

3.3.2 Pre-Treatment Procedures

After stabilization, samples have to be treated in a controlled process to quantitatively extract metals from matrices.

3.3.2.1 *Acid Digestion*

The acid digestion method is the most common, where concentrated acids are used to oxidize organic molecules and dissolve metals. The acids employed may be HNO_3, hydrochloric (HCl), sulfuric (H_2SO_4), perchloric ($HClO_4$), etc. Used mainly for soil and sediments, aqua regia is a mixture of about one volumetric part HNO_3 to three parts HCl. A nitric acid-hydrogen peroxide ($HNO_{3-}H_2O_2$) mixture is predominantly used for biological samples [16].

- ***Open-Vessel Digestion:*** Performed using hot plates; susceptible to volatilization losses and contamination.
- ***Closed-Vessel Microwave-Assisted Digestion:*** Provides improved control, lower analyte loss, less processing time, and improved safety.

3.3.2.2 *Alkaline Leaching and Fusion*

Alkaline fusion using sodium carbonate (Na_2CO_3) or lithium metaborate facilitates metal solubilization from refractory matrices (e.g., metallurgical slags, silicate-rich sediments). Leaching methodologies like SPLP (Synthetic Precipitation Leaching Procedure) and TCLP (Toxicity

Characteristic Leaching Procedure) try to mimic the scenarios where environmental release takes place [16].

3.3.2.3 Solid Phase Extraction (SPE) and Pre-Concentration

Ultra-trace analysis of aqueous samples often demands pre-concentration. Solid phase extraction (SPE) involves percolation of large volumes of samples over sorbents (e.g., iminodiacetic acid-immobilized resins, NMs-based adsorbents) to selectively capture target metals and bind them, and then elute with tiny volumes of solvents for analysis [17].

3.3.2.4 Filtration and Centrifugation

Physical separation before digestion segregates particulate-bound and dissolved fractions essential for bioavailability and speciation determination.

3.4 Analytical Methods

The ultra-trace levels at which HMs naturally exist and their speciation necessitate highly sensitive, precise, and reproducible analytical methods. With the elapse of time, trace metal analysis is significantly enhanced by spectroscopy, electrochemistry, and instrumental combination techniques to detect from nanogram to picogram per liter. Table 3.2 illustrates the principles, equipment, merits, and demerits of the most important analysis techniques employed for the detection of HMs.

3.4.1 Atomic Absorption Spectroscopy (AAS)

Atomic absorption spectroscopy (AAS) is a highly convenient and low-cost technique that takes advantage of the fact that metal atoms in the ground state will absorb a specific wavelength of radiation. The aqueous sample is taken out into the flame or into a graphite furnace, where it will be atomized. For example, Kassim *et al.* [18] in their recent report determined the quantity of copper (Cu), iron (Fe), Cr, nickel (Ni), and Pb, with a limit of detection (LOD) of 0.0142, 0.0113, 0.0641, 0.0423, and 0.1970 mg/L, respectively, in wastewater plant effluent and lake so that proper treatment can be implemented prior to its discharge. Flame AAS (FAAS) is appropriate for higher concentration levels (ppm level), and Graphite Furnace AAS (GFAAS) significantly enhances the sensitivity (ppb to ppt level) employing an electrically heated graphite tube and thus enables trace

Table 3.2 Details of various heavy metal (HM) detection and monitoring techniques.

S. no.	Techniques	Principle	Detection limit	Merits	Limitations	References
1	Atomic Absorption Spectroscopy (AAS)	Atomic Absorption Spectroscopy (AAS) uses the light absorbed by ground-state atoms to thermally emit and atomize the sample. The light absorbed is proportional to the elemental concentration of the sample following the Beer–Lambert law.	0.1 to 10.0 μg/L	High specificity, low cost per sample, well-established method	Single-element analysis requires sample digestion, limited dynamic range	[18]
2	Graphite Furnace Atomic Absorption Spectroscopy (GF-AAS)	In Graphite Furnace Atomic Absorption Spectroscopy, ultra-trace amounts of metals are quantified down to sub-ppb levels. This is achieved with the help of a specially designed electrically heated graphite tube that serves as a high-temperature atomization chamber. It generates highly efficient microliter sample atomization.	0.01 to 1.0 μg/L	Ultra-trace detection, small sample volume required	Slower than Atomic Absorption Spectroscopy (AAS), matrix interference issues	[19]

(Continued)

Table 3.2 Details of various heavy metal (HM) detection and monitoring techniques. *(Continued)*

S. no.	Techniques	Principle	Detection limit	Merits	Limitations	References
3	Inductively Coupled Plasma Mass Spectrometry (ICP-MS)	In Inductively Coupled Plasma-Mass Spectrometry (ICP–MS), the sample is ionized in a contained and sustained argon plasma. The sample is then subjected to high-vacuum mass analysis to separate by mass-to-charge ratio. Multi-elemental detection is performed down to parts-per-trillion (ppt) levels.	0.001 to 0.1 μg/L	Multi-element detection, extremely low detection limits, and a wide linear range	Expensive instrumentation and complex matrix effects require skilled operation	[21–24]
4	Inductively Coupled Plasma Optical Emission Spectroscopy (ICP-OES)	In Inductively Coupled Plasma Optical Emission Spectroscopy (ICP–OES), analyte atoms and ions are first ionized and then excited in argon plasma, which allows them to emit element-specific photons. These photons are then optically separated for multi-metal quantitation, and the process is done with high precision simultaneously.	1.0 to 10.0 μg/L	Simultaneous multi-element analysis tolerates complex matrices and fast analysis	Higher detection limits than Inductively Coupled Plasma Mass Spectrometry (ICP-MS), with spectral interferences possible	[20]

(Continued)

Table 3.2 Details of various heavy metal (HM) detection and monitoring techniques. (*Continued*)

S. no.	Techniques	Principle	Detection limit	Merits	Limitations	References
5	X-Ray Fluorescence (XRF)	X-Ray Fluorescence (XRF) is focused on the detection of high-energy X-rays that interact with a sample to excite atoms within. This detection allows for the evaluation of elemental composition by measuring secondary X-rays that are produced as a result of the high-energy incident X-rays.	1.0 to 100.0 mg/kg	Non-destructive, minimal sample preparation, portable options available	Limited sensitivity for trace levels, matrix effects are significant	[25–27]
6	Anodic Stripping Voltammetry (ASV)	Anodic Stripping Voltammetry (ASV) uses a different approach where metal ion concentration is determined through a two-step process. First, these ions are deposited onto an electrode surface by electrochemical reduction. Then, the metal is stripped off in an oxidative manner, and the resultant current peaks are linear to metal concentration.	0.01 to 1.0 μg/L	Excellent sensitivity for lead (Pb), cadmium (Cd), mercury (Hg), etc., portable field instruments available	Limited to electroactive metals; electrode fouling is possible	[28]

(*Continued*)

Table 3.2 Details of various heavy metal (HM) detection and monitoring techniques. (*Continued*)

S. no.	Techniques	Principle	Detection limit	Merits	Limitations	References
7	Cold Vapor Atomic Absorption Spectroscopy (CV-AAS)	Cold Vapor Atomic Absorption Spectroscopy (CV-AAS) selectively atomizes mercury (Hg) species to vapor mercury (Hg^0), which is carried into an optical cell for atomic absorption measurement (at 253.7 nm), allowing very sensitive detection of mercury (Hg) down to the nanogram range.	0.01 µg/L	Highly selective for mercury (Hg), with very low detection limits	Only applicable to mercury (Hg), requires careful handling	[18]
8	Hydride Generation Atomic Absorption Spectroscopy (HG-AAS)	Hydride Generation Atomic Absorption Spectroscopy (HG-AAS) reduces the elements forming hydride chemically to volatile hydrides, which, on thermal decomposition in a hot cell, are subjected to extremely selective and sensitive atomic absorption analysis.	0.1 to 1.0 µg/L	Enhanced sensitivity for hydride-forming elements reduces matrix effects	Limited to specific elements, interferences from transition metals	

(*Continued*)

Table 3.2 Details of various heavy metal (HM) detection and monitoring techniques. (*Continued*)

S. no.	Techniques	Principle	Detection limit	Merits	Limitations	References
9	Ion Chromatography (IC)	Ion Chromatography (IC) is employed to separate metal ions and their complexes in solution by ionic interaction with a stationary phase and then detected by conductivity or spectrophotometry for quantitative analysis.	0.1 to 10.0 μg/L	Speciation capability (e.g., Cr (III) vs. Cr (VI)), no need for sample digestion	Limited to ionic species, requires careful column maintenance	[44]
10	Colorimetric methods	Colorimetric assays are based on the coordination of metal ions with ligands to produce characteristic chromophores whose intensity is measured spectrophotometrically, from which metal concentrations are derived.	1.0 to 100.0 μg/L	Low-cost, field-deployable (test kits), no advanced instruments needed	Low sensitivity, subject to interference, qualitative/ semi-quantitative	[39]
11	Biosensors	Biosensors couple biological recognition components (enzymes, nucleic acids, proteins, or whole cells) to transducers that transform selective metal–biomolecule interactions into measurable optical, electrical, or electrochemical signals.	0.1 to 10.0 μg/L	Rapid, real-time monitoring, potential for field use, low sample volume	Limited stability, cross-reactivity issues, and requires calibration	[30]

(*Continued*)

Table 3.2 Details of various heavy metal (HM) detection and monitoring techniques. *(Continued)*

S. no.	Techniques	Principle	Detection limit	Merits	Limitations	References
12	Laser-Induced Breakdown Spectroscopy (LIBS)	Laser-Induced Breakdown Spectroscopy (LIBS) directs a pulsed high-power laser beam onto the surface of the sample in order to produce a microplasma, and atomic and ionic spectra obtained therefrom are used in real-time for quick, multi-element analysis.	1.0 to 100.0 mg/kg	Minimal sample preparation, simultaneous multi-element detection, portable systems available	Semi-quantitative without standards, matrix effects are significant	[45]

analysis. GFAAS accepts tiny amounts of samples and is especially suited for metals such as Pb, Cd, and Cr. For instance, Mendila and group [19] employed FAAS and GFAAS to determine concentration of Fe, manganese (Mn), zinc (Zn), Cr, Pb, Cu, Ni, and Cd) in moss and lichen samples (bio-indicator species) obtained from industrial area after wet and microwave digestion to understand the rate of anthropogenic pollution occurring in the periphery of a typical industrial zone. Despite this limitation, its accuracy and linearity make it best suited for routine laboratory environmental monitoring.

3.4.2 Inductively Coupled Plasma Optical Emission Spectroscopy (ICP-OES)

In Inductively Coupled Plasma Optical Emission Spectroscopy (ICP-OES), an inductively coupled plasma source excites atoms and ions so they may emit at element-selective wavelengths. It is particularly useful for multielement determination and routinely detects levels in the parts per billion (ppb) range. The huge capability of the ICP source to function above 6000 K is attributed to effective excitation and atomization, reducing matrix effects to the maximum extent practicable and enabling quantitative analysis in extremely complex matrices. ICP-OES is very efficient, especially for simultaneous multielement determination of a huge number of elements, and the dynamic range is enormously increased compared to the case with AAS [20]. Its two main disadvantages are its enormous operating and capital expenses and its need for argon (Ar) gas and experienced operators. Its throughput, however, sensitivity, and spectral resolution make it a first-choice instrument for high-throughput analytical laboratories.

3.4.3 Inductively Coupled Plasma Mass Spectrometry (ICP-MS)

Inductively Coupled Plasma Mass Spectrometry (ICP-MS) provides trace to ultratrace metal examination by combining ICP's ability to element-ionize with MS's ability to detect a mass-to-charge ratio. It is capable of detecting as little as parts per trillion (ppt) or, even better, with isotopic resolution, providing a discriminating power for determining metal sources and studying metal speciation. The method is of utmost importance for environmental monitoring, especially for trace metals. Moreover, it is capable of differentiating isotopes [21] (e.g., ^{206}Pb, ^{207}Pb, ^{208}Pb). Suo *et al.* [21] determined trace amounts of HM ions in water samples using silica-coated magnetic graphene oxide composites with a detection range as low as 2.023 to 13.810 ng/L. For example, this technique effectively detected the content of HMs in a total of 44 rainwater-harvested drinking water samples collected from Hebron (south of the West Bank in Palestine) [22]. Similarly, this technique is widely applied in toxicological studies and biomonitoring. For instance, HM bioaccumulation was analyzed from muscle and liver from six fish species (chub, rudd, prussian carp, common carp, brown trout, and gray mullet) collected at various locations of Neretva river using ICP-MS, which reveals presence of average 0.050 to 0.401, 0.055 to 0.703, 0.013 to 0.055, and 0.068 to 16.059 mg/kg for Hg, Pb, Cd, and Cu, respectively [23]. Although the method is unmatched in sensitivity and variety, polyatomic interferences can occur, but modern instruments integrate a Dynamic reaction cell (DRC) or collision reaction cell (CRC) to suppress them [24]. The equipment is too costly, and precise sample preparation methods require the infrastructure and advanced technical knowledge.

3.4.4 X-Ray Fluorescence (XRF) Spectroscopy

X-Ray Fluorescence (XRF) spectroscopy is considered a non-damaging technique of analysis that measures characteristic secondary (fluorescent) X-rays released upon excitation of a material with an incident primary X-ray source. XRF is commonly used for quick screening and *in situ* elemental analysis of HMs in solid waste, sediments, and soils [25]. Handheld portable XRF has enabled *in situ* HM analysis, particularly in industry and contaminated sites. The technique is well-suited to low sample preparation, non-destructive testing, and real-time results generation [25]. It is a useful tool to quantify HM levels in various samples. For example, in a recent work, Pb was quantified in raw vegetable tissues with detection limits of 0.3 μg/g employing XRF spectroscopy by simply wet coarsely-homogenizing the samples [26]. Similarly, this technique has been proven to be useful as a rapid screening technology for fast, non-destructive multi-element concentration quantification in various soil, water, and biological samples [27]. Detection limits are generally above AAS or ICP-MS, and hence the technique is best suited for preliminary surveys or bulk concentration determination. Matrix effects, surface heterogeneity, and spectral overlaps are routine problems to be dealt with by calibration using certified reference materials and applicable correction algorithms.

3.4.5 Electrochemical Techniques

Electroanalytical techniques such as Potentiometry and Anodic Stripping Voltammetry (ASV) are also rapidly surfacing for inexpensive on-site determination of HMs with high sensitivity. ASV involves preconcentration of metal ions onto the surface of an electrode, stripping by electrochemical stripping, and detection by current measurement, and hence detection at the low ppb or even ppt level. These are highly favorable to miniaturized mobile platforms and decentralized monitoring with low power consumption and miniaturization. Nanostructure electrode preparation (gold nanoparticle (Au-NPs), carbon nanotube (CNT), graphene) has introduced significantly improved sensitivity, selectivity, and stability. NM based electrochemical sensors with integration of electrochemical devices in automatic fluidic structures are currently used extensively for the detection of HMs in real water [28]. For instance, Lu *et al.* [29] reported detection of Hg^{2+}, Pb^{2+}, Cd^{2+}, and Cu^{2+} in river water along with leaching in soil and vegetables with high precision using graphene aerogel and metal–organic framework (MOF) UiO-66-NH_2 (HM binding site) composite. Electrochemical techniques are typically plagued by interference from interfering redox potentials, surfactants, and organic molecules. Electrode fouling and reproducibility are aspects that need standardization for everyday use.

3.4.6 Ultraviolet–Visible (UV–Vis) Spectrophotometry

Ultraviolet–visible (UV–Vis) spectrophotometry, while classically less sensitive than modern spectrometric equipment, remains a valuable technique of HM detection due to the ease of operation, low cost, and universal availability of the technique. The technique is based on Lambert–Beer's law principle, where chromogenic ligands such as dithizone for Pb^{2+} or Eriochrome Black T for Cr species produce a color characteristic which has an absorbance maximum, allowing rapid quantitation at the parts per million (ppm) level [16]. It is therefore perfectly suited for primary screening in low-resource laboratories and educational purposes. However, its relatively high detection limits and spectral interferences limit its use for regulation-level monitoring.

Recent developments are now overcoming such constraints. Utilization of nanostructured chromogenic ligands, functionalized NPs, and supramolecular assemblies has significantly increased sensitivity and selectivity, making possible the detection of as low as ppb levels. In addition, its integration with chemometric algorithms, digital signal processing, and miniaturized microfluidic devices has rendered it a portable, semi-automatic platform that is ideal for on-site monitoring [17]. These developments open up the applications of UV–Vis from an initial diagnostic to a more tangible analytical tool, one able to span the divide between inexpensive colorimetry and expensive spectrometric equipment, making it more feasible for decentralized and sustainable environment-quality analysis.

3.5 Nanotechnology (NT) and Biosensing-Based Methods

The aforementioned sensitive methods for the traditional analysis come with such constraints as high operational expenses, handling complexity, or the necessity of a high-grade infrastructure. On the other hand, biosensing techniques, especially when merged with NMs, serve as a perfect alternative for cheap, rapid, and sensitive detections that can be performed on-site in real time. With the coupling of biology and nanotech, new paradigms of sensing have developed that take advantage of molecular specificities and nanoscale signal amplification to allow trace levels of metal ions to be determined at ultratrace concentrations.

3.5.1 Principles of Biosensing for Heavy Metals (HMs)

Biosensors are analytical devices that are composed of an interface between a physicochemical transducer and a biological recognition element that provides a quantifiable or measurable response to the presence of a known analyte. The recognition element can be enzymes, DNA/ribonucleic acid (RNA) aptamers, peptides, whole cells, or metalloregulatory proteins, depending on the specificities and binding affinities toward certain metal ions such as Pb^{2+}, Cd^{2+}, Hg^{2+}, or As^{3+} [30]. These bio-moiety molecules respond by showing conformational or redox changes in metal complexation that are transduced into optical, electrochemical, piezoelectric, or thermal signals. These include electrochemical and optical biosensors, which are specifically ubiquitous because they have acceptable sensitivity, miniaturizability, and portability with portable instrumentation.

3.5.2 Nanomaterials (NMs) as Signal Amplifiers and Transducers

Nanomaterials (NMs) have entered into and revolutionized biosensing for signal amplification and a larger surface area for attachment of the biorecognition element. Some of the widely used nanostructures include metal NPs, quantum dots (QDs), graphene, CNTs, and MOFs. Their electronic and optical properties, like Localized Surface Plasmon Resonance (LSPR), fluorescence quenching, and electron mobility, provide excellent amplification in the sensor response. For example, Au-NPs with a DNA aptamer constitute the platform for colorimetric sensors whose color is modified due to target metal binding-mediated NP agglomeration. The DNA-AuNP conjugate demonstrated dual-channel multidimensional sensing of 9 HM ions (silver (Ag^{+}), Hg^{2+}, Cr^{3+}, tin (Sn^{4+}), Cd^{2+}, Cu^{2+}, Pb^{2+}, Zn^{2+}, and Mn^{2+}) with a detection limit of 50 nM [31]. The distinguished colors with each metal, depending on the resulting aptamer, have been illustrated in Figure 3.1. HMs such as Hg^{2+} also quench QDs *via* electron transfer, allowing for fluorescent quantification. Graphene oxide and CNTs are superior electron conductors as well as biocompatible and thus utilized in electrochemical biosensors [29] in which alteration in current is proportionally differential to HM concentration.

3.5.3 Electrochemical Biosensors

Electrochemical biosensors are among the most practical platforms for HM analysis because of their relative ease of fabrication, portability, and high sensitivity. These biosensors convert biological interactions into electrical signals, typically using amperometry, voltammetry, or impedimetry. Nanostructured electrodes, such as graphene-modified glassy carbon, AuNP-decorated screen-printed electrodes, or zinc oxide (ZnO) nanowire arrays, greatly reduce electron transfer kinetics and surface loading of recognition elements [32]. DNA electrochemical biosensors have also been extensively studied; when binding occurs or HM-induced DNA damage happens, it causes changes in charge transfer resistance or peak current, allowing for quantitative detection. New advances include label-free impedimetric aptasensors for the detection of picomolar levels of Pb^{2+} on nanostructured electrodes. Biomimetic sensors of metallothioneins or constructed bacteria are biologically meaningful sensing systems for the imitation of natural metal regulation [32].

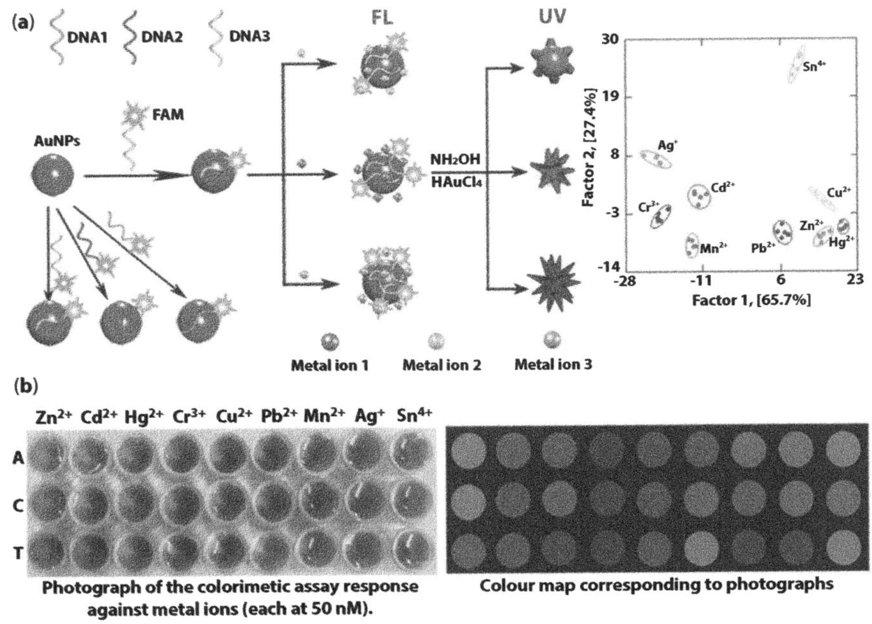

Figure 3.1 (a) Schematic of a dual-channel sensor array showing both the colorimetric signal from *in situ* grown gold-nanoparticles (Au-NPs) and fluorescence quenching using deoxyribonucleic acid-gold nanoparticle (DNA–AuNP) conjugates. (b) Visual colorimetric responses of the assay against different metal ions and corresponding color map representation. Reprinted from [31], Copyright (2016) with permission from Elsevier (License Number: 6094611093700).

3.5.4 Optical Biosensors

Optical biosensors utilize light-induced effects like absorbance, fluorescence, or surface plasmon resonance to transduce the occurrence of biomolecular interactions. Fluorescent biosensors utilizing NMs like QDs or carbon dots can offer low detection limits and sensitivity. Förster Resonance Energy Transfer (FRET) based structures are especially well-suited for HM quantitation due to their sensitivity towards molecular distance and conformational change with analyte binding [33]. AuNP aggregation colorimetric biosensors continue to find applications because of their simplicity and detectability by the naked eye. In a basic system, target metal-free aptamer-functionalized AuNPs are dispersed, but with metal binding, aggregate and turn color, cross-linking particles. The approach is compatible with fast screening of water samples without any equipment [31]. Surface-enhanced Raman scattering (SERS) is a new optical method using plasmonic nanostructures to enhance Raman signals of metal-analyte interaction with femtomolar sensitivity. SERS sensors have been used for the finding of Hg^{2+} and As^{3+} on thiolated ligand-functionalized Ag nanorod substrates [34].

3.5.5 Whole-Cell and Enzyme-Based Biosensors

Whole-cell biosensors are intact living microorganisms genetically engineered to induce a detectable response, like luminescence, fluorescence, or color change, to toxic metal ions on exposure. Most prevalent methods involve the utilization of metal-sensitive promoters together with reporter genes (e.g., lux, gfp, lacZ), resulting in biologically integrated sensing [35]. Though more ecologically relevant and less expensive, issues relating to genetic stability and sensitivity towards environmental perturbation continue to exist. Enzyme-based biosensors utilize the inhibition or activation of specific enzymes when they bind to HMs. Urease and acetylcholinesterase, for example, are widely used in sensors for Pb^{2+} and Hg^{2+} as the metals suppress the enzymic reaction, which is measurable electrochemically or by spectrophotometry. Kumar *et al.* [36] demonstrated fluorimetric detection of Pb^{2+} and Hg^{2+} with a limit of detection of 1.47×10^{-14} M (Hg^{2+}) and 0.324×10^{-10} M (Pb^{2+}), using Ag-NPs stabilized with thiocholine (TCh), which was obtained from hydrolysis of acetylcholine [36].

3.6 Remote Sensing and Spectroscopic Monitoring

Heavy metal (HM) detection and quantification in environmental matrices must be done with analytical methods that are sensitive, selective, encompass space, and respond rapidly. Remote sensing and spectroscopic monitoring have therefore become an imperative tool, offering multidimensionality on metal contamination by the contact of electromagnetic radiation with matter. Such methods comprise satellite and airborne hyperspectral imaging (HSI), handheld spectrophotometry, and colorimetric and fluorometric principle-based optical sensors. A detailed scientific description of these methods with emphasis on their physicochemical bases, analytical capabilities, and application settings is presented in this chapter.

3.6.1 Fundamentals of Remote Sensing in Heavy Metal (HM) Assessment

Remote sensing is the process of recording electromagnetic radiation reflected, emitted, or absorbed by the surface in trying to deduce compositional and structural information without physical contact. The HMs themselves are not optically active in the visible range, but their occurrence has very significant indirect influences on the spectral signatures in soils, sediments, and vegetation [37]. HM contamination has a propensity to cause physiological stress in plants that impacts chlorophyll composition, carotenoid content, and cellular organization. These biophysical and biochemical changes manifest as changes in vegetation reflectance spectra, namely in the near-infrared (700 to 1300 nm) and visible (400 to 700 nm) bands. For example, metal-induced chlorosis lowers the blue and red light absorption and raises the reflectance of the green wavelength, and this effect can be observed quantitatively by the vegetation indices such as Red Edge Position and Normalized Difference Vegetation Index (NDVI) [37]. HSI widens these capabilities to measure neighboring narrow spectral bands (usually 5 to 10 nm bandwidth) in broad spectral ranges (400 to 2500 nm). The spectral resolution is high enough to identify small spectral features, such as small metal-organic complexes or metal oxide absorption bands too small for multispectral sensors to detect. Methods including continuum removal, derivative spectroscopy, and spectral angle mapping are some of the ways in which signals of HM contamination may be disentangled from variation in the background [37].

In addition, remote sensing platforms employ thermal infrared and microwave bands to estimate soil structural parameters, moisture, and temperature, which are indirectly related to metal mobility and bioavailability. Integration of multisensor data supports predictive modeling of contamination distribution and risk evaluation. Machine learning (ML) and chemometric techniques such as Support Vector Machines (SVM), Partial Least Squares Regression (PLSR), and Random Forests (RF) have now been positioned to take a central role in the interpretation of high-dimensional hyperspectral data to enable the classification and quantification of contaminated sites with greater accuracy and robustness [38].

3.6.2 Spectroscopic Monitoring: Principles and Modalities

Spectroscopic sensing depends on the intrinsic interaction between the molecules and electromagnetic radiation in the form of absorption, emission, scattering, or fluorescence. These interactions create molecular fingerprints that can be used for sensitive direct determination of HMs or *via* specific probes.

3.6.2.1 Colorimetric Sensing

Colorimetric detection is based on the development of chromophoric complexes between the HM ions and discriminating ligands with shared changes of optical absorption. The Beer–Lambert law regulates the quantitative correlation between the analyte concentration and the absorbance that permits direct calibration and quantification. Dithizone, 4-(2-pyridylazo) resorcinol (PAR), and 1-(2-pyridylazo)-2-naphthol (PAN) are broadly used as the reagents for Pb (II), Hg (II), and Cd (II) metals, which give colored chelates with well-separated absorption maxima (for example, dithizone-Pb chelate absorbs around 510 nm). Assay sensitivity is influenced by ligand binding constants (log K), participation rate in the reaction, and solution conditions like pH and ionic strength. Azmi and Low [39] utilized diphenylthiocarbazone (dithizone, DTZ), which was immobilized onto cellulose acetate, to obtain stable, long-term membranes with naked-eye sensing capabilities for Pb^{2+}, Cu^{2+}, Cd^{2+}, and Hg^{2+}. The addition of donor-rich -OH and -NH functionalities allowed for strong coordination with target ions, showing high stability (90 days) as well as practical field applicability [39].

Current developments include nanoscale materials with LSPR, where metal NP conduction electrons oscillate in response to incident light. Thiol- or amine-functional terminatable Au and Ag NPs respond with size- and aggregation-dependent coloration. With coordination of HMs upon NP aggregation, the LSPR band is shifted from ~520 nm (red) to longer wavelengths (~600 to 700 nm, blue/purple), and this can be read spectrophotometrically or visually [31]. This plasmonic colorimetric sensor technique has ultra-low detection limits (ppm) and fast response times (order of seconds to minutes). Selectivity is increased through functionalization with metal-specific aptamers, peptides, or molecularly imprinted polymers with discrimination between chemically related ions. Si *et al.* [40] reported a polyaniline (PANI) based colorimetric membrane sensor for Hg^{2+} detection showing a chromatic change distinguishable from white–yellow to green–blue, with a visual detection limit of 5 nM. Significantly, the membrane was considered reusable since it showed remarkable reversibility and stability over several regeneration cycles. Likewise, Hu *et al.* [34] fabricated a dual-mode sensor by combining Ag nanocrystals with 3,3′,5,5′-tetramethylbenzidine (TMB)-modified cellulose nanofibers (CNFs) for colorimetric and

SERS-based detection. While the colorimetric mode had a detection limit of ~33 nM, the SERS one could reach an ultratrace sensitivity of ~1.1 nM.

Analytical methodologies increasingly favor smartphone-based colorimetric analysis with digital image analysis. Derived Red, Green, and Blue (RGB) values can be quantitatively translated to concentration measures from calibration curves to produce decentralized and real-time monitoring with low instrumentation. Sample matrix coloration, turbidity, and pH changes introduce inherent problems of colorimetric sensing that impinge on ligand stability and spectral characteristics. Immobilization of colorimetric reagents on solid substrates such as paper-based microfluidics or hydrogel matrices mitigates these and enhances reproducibility and convenience.

3.6.2.2 Fluorometric Sensing

Fluorometric sensing is generally more selective and sensitive than colorimetry because it takes advantage of the photophysical processes behind fluorescence emission upon electronic excitation. The intensity, lifetime, and spectral position of the fluorescence may be manipulated by HM binding through FRET, intramolecular charge transfer (ICT), photoinduced electron transfer (PET), heavy atom effect, or aggregation-induced emission (AIE). Synthesis of fluorometric probes entails the conjugation of the chelating groups with high specificity and affinity for the target metal ions to fluorophores. Conformational alterations in the probe's molecular structure upon binding affect electron distribution, leading to "turn-on," "turn-off," or ratiometric fluorescent responses [41].

Förster resonance energy transfer (FRET), originally described by Theodor Förster in 1948, is a non-radiative dipole–dipole contact in which an excited donor fluorophore transfers its energy to an adjacent acceptor within the Förster distance, usually 1 to 10 nm [42]. The acceptor can be a fluorophore or a non-emissive chromophore, and transfer efficiency relies heavily on spectral overlap and spatial proximity. In contrast, PET is charge separation on photoexcitation, where the donor and acceptor result in radical ion pairs that undergo non-radiative recombination [12]. PET is strongly distance-dependent (<10 Å) and may be modulated through electron-withdrawing substituents on the chelating unit [33]. Inner filter effects (IFE) happen when the analyte absorbs and spectrally overlaps with the fluorophore, resulting in apparent quenching but not lifetime diminishment [33].

Quantum dots (QDs), size-tunable nanoscale semiconductor particles, have transformed HM detection. Thiol or carboxyl ligand functionalization of QDs allows selective chelation of metals. Coordination of the metal ions changes the surface state of the QD, leading to fluorescence quenching by non-radiative recombination or sensitization by passivation. Large-absorbing, photostable, highly quantum-yield-emitting, narrow-emission-spectrum QDs are unparalleled in their sensitivity to picomolar concentrations. Oxygen (O_2) and nitrogen (N) rich moieties enriched carbon dots have proved to be potent PET and FRET-based HM sensors. For example, nitrogenous carbon dots loaded natural mineral (microcline) based free-standing and insoluble membrane could perform dual-sensing of Cr^{6+} and Fe^{3+} through PET and spectral red-shift mechanisms [33], highlighting translational potential for real-world environmental monitoring (Figures 3.2a-b). Notably, Roy *et al.* [43] designed a boehmite–CD/polymer nanocomposite-based handheld portable "Kavach" probe, functional in all pH values, to detect Cr^{6+} in wastewater, as illustrated in Figure 3.2c. Heteroatom doping also boosts quantum yield, allowing strong, flexible films with good photostability.

Current advances drive fluorometric probes towards miniature, miniaturized platforms with light-emitting diode (LED) excitation and photodiode detection. Such devices allow point-of-care environmental monitoring with fast readout and minimal sample pretreatment. Fluorometric sensors are subject to photobleaching, autofluorescence interference by environmental organic molecules, and the complexity of probe synthesis, while offering advantages. Development focuses on robust, photostable fluorophores with increased metal ion selectivity and sensitivity for complex sample matrices.

3.7 Limitations and Challenges in Detection

Despite the enormous growth in analytical and sensor technology used to detect HMs, there are several inherent and operational issues that still exist that detract from the field-translatability, reliability, and accuracy of current methodologies. The limitations are due to complex environmental matrices, metal speciation kinetics, limitations in equipment, and the requirements for low-cost, sensitive, and selective field-deployable processes.

3.7.1 Matrix Complexity and Interferences

Environmental matrices, soil, and water to biota are complex in composition with numerous inorganic ions, natural organic compounds, and particulate interferents which influence quantitation of metals. Matrix

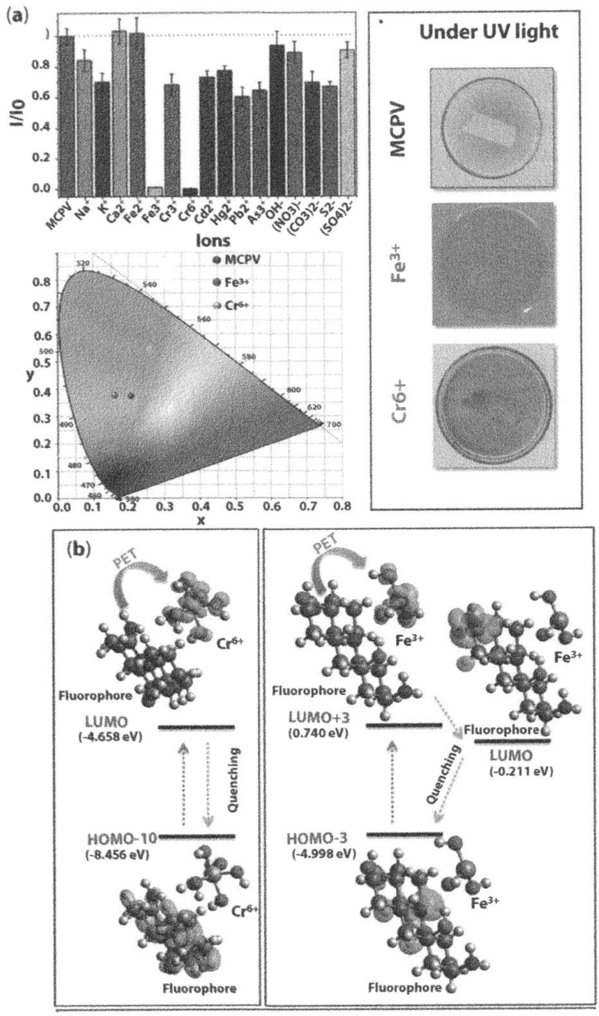

Figure 3.2 (a) Demonstration of dual detection of chromium (Cr^{6+}) and iron (Fe^{3+}) ions, with naked-eye observation under ultraviolet (UV) illumination. (b) Photoinduced electron transfer (PET) quenching mechanism illustrating molecular orbitals and electronic transitions when the sensor comes in contact with quencher ions. Reprinted from [33], Copyright (2021) with permission from Springer Nature. (License Number: 6094620992834); (c) Portable, reusable sensing device (Kavach) for Cr^{6+} fabricated using a carbon dot/boehmite/Polyvinylidene Difluoride (PVDF) membrane. Reprinted from [43], Copyright (2021) with permission from Elsevier (License Number: 6094621247885).

(Continued)

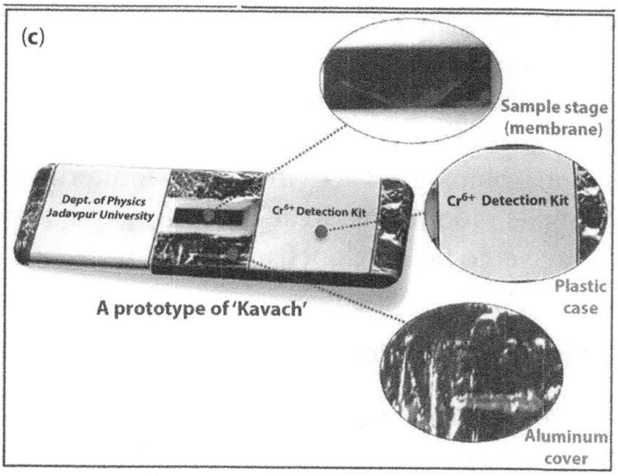

Figure 3.2 (Continued) (a) Demonstration of dual detection of chromium (Cr^{6+}) and iron (Fe^{3+}) ions, with naked-eye observation under ultraviolet (UV) illumination. (b) Photoinduced electron transfer (PET) quenching mechanism illustrating molecular orbitals and electronic transitions when the sensor comes in contact with quencher ions. Reprinted from [33], Copyright (2021) with permission from Springer Nature. (License Number: 6094620992834); (c) Portable, reusable sensing device (Kavach) for Cr^{6+} fabricated using a carbon dot/boehmite/Polyvinylidene Difluoride (PVDF) membrane. Reprinted from [43], Copyright (2021) with permission from Elsevier (License Number: 6094621247885).

effects dilute analytical signals, especially in spectroscopic analysis such as colorimetric and fluorometric sensing, where turbidity, pH change, and coloring of the sample can interfere with optical determination and cause false positives. Coordination of metal ions by fulvic and humic acids or ionic strength variation alters the availability of metals and makes detection more difficult. Development of matricolerant sensors using selective membranes, aptamers, or molecularly imprinted polymers is of significant research interest [46].

3.7.2 Speciation and Bioavailability

The total metal content gives little insight into toxicity and environmental mobility. Metals exist in several states of chemistry, free ions, complexes, organometallics, or bound on PM, and ecological risk is strongly a function of bioavailable fractions. Few operations, however, can distinguish oxidation states or species. Sophisticated techniques such as X-ray absorption spectroscopy and high-performance liquid chromatography-Inductively

Coupled Plasma Mass Spectrometry (HPLC-ICPMS) provide valuable information about speciation; these are capital-intensive, instrument-intensive, and are unsuitable for a routine or field-level approach [47]. The physicochemical background keeps changing with respect to pH, redox potentials, and microbiological activities, further altering metal speciation and complicating its monitoring. Real-time speciating and discriminating *in situ* sensors are yet to be developed.

3.7.3 Sensitivity and Detection Limits

Trace metal level measurements, typically less than ppb, are required to meet regulatory safety requirements. While highly developed laboratory techniques, such as ICP-MS or AAS, measure at sub-ppb sensitivity, hand-held sensors show high detection levels and low dynamic ranges. Environmental interference and sample complexity require pre-concentration or signal amplification methods, which introduce procedural complexity. Moreover, the ultra-sensitivity against sensor robustness trade-off remains a technical challenge as highly sensitive probes will rapidly destabilize or cross-react, but more robust sensors compromise detection resolution [48].

3.7.4 Selectivity and Cross-Reactivity

Lead (Pb^{2+}), Cd^{2+}, and Zn^{2+} modes of coordination chemistries preclude the realization of selective detection. Ligand- or thiol-based sensors bind to more than one metal, and optical sensors are deceived by colored molecules or interfering ions with cross-absorbing spectra. Methods such as aptamer design, molecular imprinting, and multi-modal detection attempt to improve selectivity, but complete specificity is beyond them [48].

3.7.5 Instrumentation and Field Deployment

Being restricted by size, power draw, and environmental interference, advanced field analytical instruments (e.g., ICP-MS, LIBS, XRF) are only suitable for laboratory purposes. Hand-held instruments also suffer limitations of calibration, data drift, cause environmental instability, and have low power sources. Remote sensing technologies, although promising, are constrained by low spatial resolution and complexity in analysis and ground validation [48].

3.7.6 Standardization and Ethical Issues

Lack of established practice, especially for emerging sensor technologies, impedes intercomparability across laboratories and uptake into regulation. Quality assurance is certified reference material cross-validation. Ethics of sensor material sustainability and remote sensing data privacy remain under-addressed. Demonstrated access to technology and sustainable management of environmentally benign material are essential for the practice of sustainable monitoring [49].

3.8 Future Prospects and Conclusion

Continued global concern for HM contamination as a two-way ecosystem and human-health hazard is driving a revolution in monitoring technology that demands ultra-high sensitivity, specificity, and spatiotemporal resolution at affordable cost for productive field deployment. Conventional high-standard platforms such as AAS, ICP-MS, and XRF offer multi-element trace detection but at the expense of infrastructure requirements and unsuitability for rapid field deployment. To this, low-cost, quick electrochemical, colorimetric, and fluorometric sensors have emerged as portable devices, but matrix interference, selectivity, and stability continue to be issues in complex environmental matrices. Interfacing NT, hierarchical nanostructures, and functional materials is the solution to overcoming such obstacles through the design of reusable, multiplexed sensors with optoelectronic tunability, self-assembly, or 3-D nanoprinting for ultra-trace discrimination stability. Lab-on-chip and microfluidic integration will miniaturize sampling, reaction, and detection, enable high-throughput multi-metal analysis with low reagent consumption, and smartphone interfaces and wireless modules will make decentralized, real-time monitoring even feasible in remote areas through energy-autonomous devices.

Hyperspectral sensing from Unmanned Aerial Vehicles (UAVs) and nanosatellites, backed up by Light Detection and Ranging (LiDAR), thermal sensing, and Geographic Information System (GIS) platforms facilitated by artificial intelligence (AI) based analytics at the macro-scale, will deliver automated anomaly detection, predictive modeling, and dynamic risk mapping. Quantum sensing will continue to lower detection thresholds below classical limits, and distributed sensor networks will require robust AI/ML pipelines and edge–cloud infrastructures to self-calibrate, fuse heterogeneous data streams, and generate actionable intelligence. Equally critical is sensor sustainability, such as biodegradable platforms, eco-friendly NMs, and community-based participatory monitoring, which can de-monetize exposure surveillance. The fate of HM monitoring, therefore, is a holistic, self-sustaining, and enduring system that integrates advanced materials, sensing technology, and data knowledge to provide proactive remediation and preserve both environmental integrity and public health.

List of Abbreviations

AAS	Atomic Absorption Spectroscopy
Ag	Silver
AI	Artificial intelligence
AIE	Aggregation-induced emission
Ar	Argon
As	Arsenic
ASV	Anodic Stripping Voltammetry
Au	Gold
Au-NPs	Gold nanoparticle
Cd	Cadmium
CNFs	Cellulose nanofibers
CNT	Carbon nanotube
COPD	Chronic obstructive pulmonary disease
Cr	chromium
CRC	Collision reaction cell
Cu	Copper
CV-AAS	Cold Vapor-Atomic Absorption Spectrometry
DNA	Deoxyribonucleic acid
DRC	Dynamic reaction cell
DTZ	Diphenylthiocarbazone
EC	Electrical conductivity
FAAS	Flame Atomic Absorption Spectrometry
Fe	Iron
Fe_2O_3	Iron oxide
FRET	Förster resonance energy transfer
GF-AAS	Graphite Furnace Atomic Absorption Spectrometry
GIS	Geographic Information System
H_2O_2	Hydrogen peroxide
H_2SO_4	Sulfuric acid
HCl	Hydrochloric acid
$HClO_4$	Perchloric acid

Hg	Mercury
HG-AAS	Hydride Generation-Atomic Absorption Spectrometry
HIS	Hyperspectral imaging
HMs	Heavy metals
HNO_3	Nitric acid
HPLC-ICPMS	High-performance liquid chromatography-Inductively Coupled Plasma Mass Spectrometry
IC	Ion chromatography
ICP-MS	Inductively Coupled Plasma Mass Spectrometry
ICP-OES	Inductively Coupled Plasma - Optical Emission Spectrometry
ICT	Intramolecular charge transfer
IFE	Inner filter effects
ISO	International Organization for Standardization
LED	Light-emitting diode
LIBS	Laser-induced breakdown spectroscopy
LiDAR	Light Detection and Ranging
LOD	Limit of detection
LSPR	Localized surface plasmon resonance
ML	Machine learning
Mn	Manganese
MOF	Metal–organic framework
N	Nitrogen
Na_2CO_3	Sodium carbonate
NDVI	Normalized Difference Vegetation Index
Ni	Nickel
NM	Nanomaterial
NT	Nanotechnology
O_2	Oxygen
OSHA	Occupational Safety and Health Administration
P	Phosphorus
PAN	1-(2-pyridylazo)-2-naphthol
PANI	Polyaniline
PAR	4-(2-pyridylazo) resorcinol
Pb	Lead
PE	Polyethylene
PET	Photoinduced electron transfer
PLSR	Partial Least Squares Regression
PM	Particulate matter
ppb	Parts per billion

ppm	Parts per million
ppt	Parts per trillion
PVDF	Polyvinylidene Difluoride
QDs	Quantum dots
RF	Random Forests
RGB	Red, Green, and Blue
RNA	Ribonucleic acid
SERS	Surface-enhanced Raman scattering
Sn	Tin
SPE	Solid Phase Extraction
SPLP	Synthetic Precipitation Leaching Procedure
SVM	Support Vector Machines
TCh	Thiocholine
TCLP	Toxicity Characteristic Leaching Procedure
TMB	3,3′,5,5′-tetramethylbenzidine
UAVs	Unmanned Aerial Vehicles
UNEP	United Nations Environment Programme
USEPA	United States Environmental Protection Agency
UV-Vis	Ultraviolet–visible
WHO	World Health Organization
XRF	X-ray Fluorescence
Zn	Zinc
ZnO	Zinc oxide

AI disclosure statement: The author acknowledges the use of AI-based tools for minor language editing or grammatical correction. However, all intellectual contributions, analyses, and conclusions presented in this chapter are entirely the author's own.

References

1. Mulware, S.J., Toxicity of heavy metals, A., Subject in review. *Int. J. Recent Res. Phys. Chem. Sci.*, 6, 2, 30–43, 2020.
2. Bhardwaj, L.K., Sharma, S., Jindal, T., Estimation of physico-chemical and heavy metals in the lakes of Grovnes & Broknes Peninsula, Larsemann Hill, East Antarctica. *Chem. Afr.*, 6, 5, 2677–2694, 2023.
3. Alam, S., Bhardwaj, L.K., Mallick, R., Rai, S., Estimation of heavy metals and fluoride ion in vegetables grown nearby the stretch of River Yamuna, Delhi (NCR), India. *Indian J. Environ. Prot.*, 43, 1, 64–73, 2023.

4. Shah, S.B., Heavy metals in the marine environment—an overview, in: *Heavy Metals in Scleractinian Corals*, pp. 1–26, Springer, Cham, 2021.
5. Driscoll, C.T., Mason, R.P., Chan, H.M., Jacob, D.J., Pirrone, N., Mercury as a global pollutant: sources, pathways, and effects. *Environ. Sci. Technol.*, 47, 10, 4967–4983, 2013.
6. Pacyna, J.M., Sundseth, K., Pacyna, E.G., Sources and fluxes of harmful metals, in: *Environmental Determinants of Human Health*, pp. 1–25, Springer International Publishing, Cham, 2016.
7. Raj, K. and Das, A.P., Lead pollution: Impact on environment and human health and approach for a sustainable solution. *Environ. Chem. Ecotoxicol.*, 5, 79–85, 2023.
8. Yang, Y., Hassan, M.F., Ali, W., Zou, H., Liu, Z., Ma, Y., Effects of cadmium pollution on human health: A narrative review. *Atmosphere*, 16, 2, 225, 2025.
9. Malik, L.A., Bashir, A., Qureashi, A., Pandith, A.H., Detection and removal of heavy metal ions: a review. *Environ. Chem. Lett.*, 17, 4, 1495–1521, 2019.
10. Shi, Y., Zhang, S., Zhou, H., Dong, Y., Liu, G., Ye, W., He, R., Zhao, G., Recent developments in heavy metals detection: Modified electrodes, pretreatment methods, prediction models and algorithms. *Metals*, 15, 1, 80, 2025.
11. Sevak, P. and Pushkar, B., Arsenic pollution cycle, toxicity and sustainable remediation technologies: A comprehensive review and bibliometric analysis. *J. Environ. Manage.*, 349, 119504, 2024.
12. Bardhan, S., Roy, S., Chanda, D.K., Ghosh, S., Mondal, D., Das, S., Das, S., Nitrogenous carbon dot decorated natural microcline: an ameliorative dual fluorometric probe for Fe 3+ and Cr 6+ detection. *Dalton Trans.*, 49, 30, 10554–10566, 2020.
13. Chakraborty, S. and Rayalu, S., Detection of nickel by chemo and fluoro sensing technologies. *Spectrochim. Acta, Part A*, 245, 118915, 2021.
14. Elkhatat, A.M., Soliman, M., Ismail, R., Ahmed, S., Abounahia, N., Mubashir, S., Fouladi, S., Khraisheh, M., Recent trends of copper detection in water samples. *Bull. Natl. Res. Cent.*, 45, 1, 218, 2021.
15. Van, H.T. and Nga, L.T.Q., Effects of Zn pollution on soil: Pollution sources, impacts and solutions. *Results Surf. Interfaces*, 17, 100360, 2024.
16. Csuros, M., *Environmental sampling and analysis for technicians*, CRC Press, Taylor and Francis, 2018.
17. Zhang, J. and Zhang, C., Sampling and sampling strategies for environmental analysis. *Int. J. Environ. Anal. Chem.*, 92, 4, 466–478, 2012.
18. Kassim, N.A., Ghazali, S.A.I.S.M., Bohari, F.L., Abidin, N.A.Z., Assessment of heavy metals in wastewater plant effluent and lake water by using atomic absorption spectrophotometry. *Mater. Today Proc.*, 66, 3961–3964, 2022.
19. Mendila, D., Uzuna, F., Tuzena, M., Comparison of wet and microwave sample digestion method for the FAAS and GFAAS determination of heavy metal concentrations in moss and lichen samples. *At. Spectrosc.*, 38, 18–24, 2017.

20. Khan, S.R., Sharma, B., Chawla, P.A., Bhatia, R., Inductively coupled plasma optical emission spectrometry (ICP-OES): a powerful analytical technique for elemental analysis. *Food Anal. Methods*, 15, 3, 666–688, 2022.
21. Suo, L., Dong, X., Gao, X., Xu, J., Huang, Z., Ye, J., Lu, X., Zhao, L., Silica-coated magnetic graphene oxide nanocomposite based magnetic solid phase extraction of trace amounts of heavy metals in water samples prior to determination by inductively coupled plasma mass spectrometry. *Microchem. J.*, 149, 104039, 2019.
22. Malassa, H., Al-Rimawi, F., Al-Khatib, M., Al-Qutob, M., Determination of trace heavy metals in harvested rainwater used for drinking in Hebron (south West Bank, Palestine) by ICP-MS. *Environ. Monit. Assess.*, 186, 10, 6985–6992, 2014.
23. Djedjibegovic, J., Larssen, T., Skrbo, A., Marjanović, A., Sober, M., Contents of cadmium, copper, mercury and lead in fish from the Neretva river (Bosnia and Herzegovina) determined by inductively coupled plasma mass spectrometry (ICP-MS). *Food Chem.*, 131, 2, 469–476, 2012.
24. Markiewicz, B., Komorowicz, I., Barałkiewicz, D., Accurate quantification of total chromium and its speciation form Cr (VI) in water by ICP-DRC-IDMS and HPLC/ICP-DRC-IDMS. *Talanta*, 152, 489–497, 2016.
25. Pearson, D., Weindorf, D.C., Chakraborty, S., Li, B., Koch, J., Van Deventer, P., de Wet, J., Kusi, N.Y., Analysis of metal-laden water *via* portable X-ray fluorescence spectrometry. *J. Hydrol.*, 561, 267–276, 2018.
26. Byers, H.L., McHenry, L.J., Grundl, T.J., XRF techniques to quantify heavy metals in vegetables at low detection limits. *Food Chem.: X.*, 1, 100001, 2019.
27. McComb, J.Q., Rogers, C., Han, F.X., Tchounwou, P.B., Rapid screening of heavy metals and trace elements in environmental samples using portable X-ray fluorescence spectrometer, a comparative study. *Water Air Soil Pollut.*, 225, 12, 2169, 2014.
28. Waheed, A., Mansha, M., Ullah, N., Nanomaterials-based electrochemical detection of heavy metals in water: Current status, challenges and future direction. *TrAC, Trends Anal. Chem.*, 105, 37–51, 2018.
29. Lu, M., Deng, Y., Luo, Y., Lv, J., Li, T., Xu, J., Chen, S.W., Wang, J., Graphene aerogel–metal–organic framework-based electrochemical method for simultaneous detection of multiple heavy-metal ions. *Anal. Chem.*, 91, 1, 888–895, 2018.
30. Saidur, M.R., Aziz, A.A., Basirun, W.J., Recent advances in DNA-based electrochemical biosensors for heavy metal ion detection: A review. *Biosens. Bioelectron.*, 90, 125–139, 2017.
31. Tan, L., Chen, Z., Zhao, Y., Wei, X., Li, Y., Zhang, C., Wei, X., Hu, X., Dual channel sensor for detection and discrimination of heavy metal ions based on colorimetric and fluorescence response of the AuNPs-DNA conjugates. *Biosens. Bioelectron.*, 85, 414–421, 2016.

32. De Benedetto, G.E., Di Masi, S., Pennetta, A., Malitesta, C., Response surface methodology for the optimisation of electrochemical biosensors for heavy metals detection. *Biosensors*, 9, 1, 26, 2019.
33. Bardhan, S., Roy, S., Chanda, D.K., Mondal, D., Das, S., Das, S., Flexible and reusable carbon dot decorated natural microcline membrane: A futuristic probe for multiple heavy metal induced carcinogen detection. *Microchim. Acta*, 188, 4, 134, 2021.
34. Hu, F., Li, Y., Zhang, Y., Li, Y., Li, H., Ai, S., Flexible Ag NCs/CNFs film for colorimetric and SERS dual-mode ultrasensitive detection of mercury ions (II). *Vib. Spectrosc.*, 118, 103342, 2022.
35. Teo, S.C. and Wong, L.S., Whole cell-based biosensors for environmental heavy metals detection. *Annu. Res. Rev. Biol.*, 4, 17, 2663–2674, 2014.
36. Kumar, D.N., Roy, J., Alex, S.A., Chandrasekaran, N., Mukherjee, A., Spectrofluorimetric determination of Hg 2+ and Pb 2+ using acetylcholinesterase (AChE)-based formation of silver nanoparticles. *RSC Adv.*, 6, 25, 21261–21270, 2016.
37. Lovynska, V., Bayat, B., Bol, R., Moradi, S., Rahmati, M., Raj, R., Sytnyk, S., Wiche, O., Wu, B., Montzka, C., Monitoring heavy metals and metalloids in soils and vegetation by remote sensing: A review. *Remote Sens.*, 16, 17, 3221, 2024.
38. Guo, Y., Liang, Y., Deng, R., Li, J., Wang, J., Hua, Z., Tang, Y., Development and application of a new sensitivity analysis model for the remote sensing retrieval of heavy metals in water. *Heliyon*, 8, 12, 2022.
39. Azmi, N.A. and Low, S.C., Investigating film structure of membrane-based colorimetric sensor for heavy metal detection. *J. Water Process Eng.*, 15, 37–42, 2017.
40. Si, Y., Wang, X., Li, Y., Chen, K., Wang, J., Yu, J., Wang, H., Ding, B., Optimized colorimetric sensor strip for mercury (II) assay using hierarchical nanostructured conjugated polymers. *J. Mater. Chem. A*, 2, 3, 645–652, 2014.
41. Rasheed, T., Bilal, M., Nabeel, F., Iqbal, H.M., Li, C., Zhou, Y., Fluorescent sensor based models for the detection of environmentally-related toxic heavy metals. *Sci. Total Environ.*, 615, 476–485, 2018.
42. Müller, S.M., Galliardt, H., Schneider, J., Barisas, B.G., Seidel, T., Quantification of Förster resonance energy transfer by monitoring sensitized emission in living plant cells. *Front. Plant Sci.*, 4, 413, 2013.
43. Roy, S., Bardhan, S., Mondal, D., Saha, I., Roy, J., Das, S., Chanda, D.K., Karmakar, P., Das, S., Polymeric carbon dot/boehmite nanocomposite made portable sensing device (Kavach) for non-invasive and selective detection of Cr (VI) in wastewater and living cells. *Sens. Actuators, B*, 348, 130662, 2021.
44. Michalski, R., Ion chromatography applications in wastewater analysis. *Separations*, 5, 1, 16, 2018.
45. Martino, L.J., D'Angelo, C.A., Rodriguez, C.I., Diaz Pace, D.M., Rapid monitoring of heavy metals in fluvial sediments using laser-induced breakdown spectroscopy. *Rev. Mex. Fis.*, 65, 5, 440–448, 2019.

46. Mao, K., Zhang, H., Wang, Z., Cao, H., Zhang, K., Li, X., Yang, Z., Nanomaterial-based aptamer sensors for arsenic detection. *Biosens. Bioelectron.*, 148, 111785, 2020.
47. Tang, W., Shan, B., Zhang, H., Zhu, X., Li, S., Heavy metal speciation, risk, and bioavailability in the sediments of rivers with different pollution sources and intensity. *Environ. Sci. Pollut. Res.*, 23, 23, 23630–23637, 2016.
48. Kumar, M., Gogoi, A., Kumari, D., Borah, R., Das, P., Mazumder, P., Tyagi, V.K., Review of perspective, problems, challenges, and future scenario of metal contamination in the urban environment. *J. Hazard. Toxic Radioact. Waste*, 21, 4, 04017007, 2017.
49. Inobeme, A., Mathew, J.T., Jatto, E., Inobeme, J., Adetunji, C.O., Muniratu, M., Onyeachu, B.I., Adekoya, M.A., Ajai, A.I., Mann, A., Olori, E., Recent advances in instrumental techniques for heavy metal quantification. *Environ. Monit. Assess.*, 195, 4, 452, 2023.

4

Remediation Technologies: Science and Innovation

Prabhat Kumar Patel[1,2*] **and Ramagopal V.S. Uppaluri**[1]

[1]*Centre for the Environment, Indian Institute of Technology Guwahati, Guwahati, Assam, India*
[2]*School of Science, Harbin Institute of Technology Shenzhen, Shenzhen, China*

Abstract

Environmental contamination due to heavy metals (HMs) and their impact on human health are among the most critical concerns. However, traditional methods such as flotation, electrochemical, membrane filtration, chemical precipitation, and coagulation-flocculation methods for HM removal and remediation tend to be costly due to the use of non-regenerative materials. Addressing this challenge, prior art suggested adsorption processes utilizing various materials as adsorbents as an effective solution. Owing to its inherent advantages, such as stability, affordability, operational simplicity, and efficacy, the effectiveness of adsorption as a purification method has been well-established. By employing a variety of efficient adsorbents such as chitosan-based derivatives, adsorption technology enables a significant reduction of HM ion concentrations to incredibly low levels. For agricultural and industrial waste streams, ion exchange, chemisorption, micro-precipitation, and adsorption-complexation are a few of the variables that control the intricate process of metal sorption onto sorbent materials. The present study delves into the assortment of sorbent materials employed in mitigating the ions of metals in effluent and water courses, while also assessing the benefits of employing the adsorption process. The capability of these adsorbent materials has been evaluated within the scope of diverse HM elimination from both water and wastewater sources.

***Keywords*:** Contamination, heavy metal ions, sustainable solution, treatment technologies, wastewater

**Corresponding author*: pprabhat1991@gmail.com; ORCID: https://orcid.org/0000-0002-9001-8516

Prangya Rath, Laxmi Kant Bhardwaj and Amit Kumar (eds.) Heavy Metal Remediation, (83–108)

4.1 Introduction

Heavy metals (HMs) are some of the most dangerous environmental pollutants because they don't break down naturally. They get into water systems used by people through both natural processes and human activities [1, 2]. Even at low concentrations (parts per million or milligrams per liter), these metals are toxic to living organisms. Some HMs, like chromium (Cr), cadmium (Cd), arsenic (As), iron (Fe), and nickel (Ni), occur naturally in the Earth's crust [3]. However, increased industrialization and urbanization have led to an excessive build-up of these metals in the environment [4]. Effluent comes from diverse sources, including commercial, industrial, domestic, and agricultural activities, as well as surface runoff [5]. Industrial wastewater is the main source of HM pollution because many industries use high concentrations of these metals in their processes [6]. As a result, these pollutants often end up in irrigation water through industrial effluents [7]. Table 4.1 lists common commercial sources of HM-polluted wastewater and their adverse consequences on the human body.

As a tiny component, numerous HMs, including zinc (Zn), lead (Pb), and copper (Cu), may have a beneficial and synergistic effect on human metabolism and health [8]. However, they are harmful and dangerous at higher levels of pollution, and they may even cause drinking water pollution, which may infect the human body. Conversely, increased airborne HM contaminants are often significant at emission locations [9]. So, the gradual buildup of HMs in the body originates from a combination of drinking water pollution, contaminants in the air, and sediment buildup [10]. Much slower buildup eventually results in a deadly situation for all of humanity and necessitates the implementation of efficient mitigating techniques.

It is both intriguing and difficult to get rid of HM contaminants like Cu, Zn, Pb, and Fe from waste liquid systems. In the specific domain focused on the elimination of hazardous HMs from spent substances, various methodologies, including membrane separation, electrochemical reduction, ion exchange resins, nanomaterials (NMs), chemical precipitation, adsorption processes, coagulation, and flocculation, are used [11–13], have been documented for their effective performance. Solving the problem of removing HMs from water in a sustainable and cost-effective way is a major challenge. The best solutions need to be environmentally friendly, affordable, and long-lasting, and they also need to be quick and easy to research

Table 4.1 A summary of the various origins of heavy metal (HM) effluents and their detrimental effects.

S. no.	Metals	Harmful effects	Prominent sources
1	Chromium (Cr)	Respiratory problems	Tanneries, paper and pulp industries, petroleum plants, chrome plating industries, and electroplating industries
2	Iron (Fe)	Iron (Fe) overload diseases	Mining, metal refineries, and engine parts
3	Zinc (Zn)	Gastrointestinal issues and essential mineral absorption	Rubber, pigments, and paints industries, wood preservatives, and ointments
4	Mercury (Hg)	Neurological problems in children and cardiovascular issues	Thermometers, bulbs, adhesives, leather industries, paints, tanning, and ointments
5	Lead (Pb)	Developmental issues and cognitive problems in children	Batteries, pesticides, and petroleum-based industries
6	Arsenic (As)	Cardiovascular diseases, skin problems, and cancer	Automotive exhaust, dyes, and wood conservators
7	Cadmium (Cd)	Osteoporosis, kidney damage, and lung cancer	Plating, fertilizers, pigments, galvanized pipes, polyvinyl refineries, pesticides, and battery industries

(Continued)

Table 4.1 A summary of the various origins of heavy metal (HM) effluents and their detrimental effects. (*Continued*)

S. no.	Metals	Harmful effects	Prominent sources
8	Aluminum (Al)	Neurotoxicity	Pesticides, wiring products, ceramics, aluminum phosphate industries, and automobile parts
9	Copper (Cu)	Liver damage and gastrointestinal problems	Metal refining, plating industries, and plastic industries

and implement [14]. Current technologies for treating HM-contaminated wastewater generally fall into three main categories:

(i) Chemical methods use chemical reactions to remove metals. These include chemical precipitation, electrochemical reduction, and electrolysis.
(ii) Physical methods remove metals through phase changes or sorption. Examples are biomass adsorption, ion exchange resins, reverse osmosis, and evaporation.
(iii) Biological methods rely on biological systems to adsorb or enrich metals. This category includes biosorption, bioflocculation, and artificial wetlands.

4.2 Research Gaps

The elimination of HMs from polluted water bodies has served as the focus of a significant number of earlier studies as well as continuing research investigations. Among them, significant investigations focused on the elimination of HMs from simple artificial aqueous solutions, including neutral or acidic solutions. Nevertheless, real-time situations concerning commercial waste effluents exhibit increased complexity due to the presence of additional cations [15]. Even the most well-cited previous work on the topic of adsorption-desorption has failed to take such systems into account. Industrial and agricultural unprocessed wastewater is an example of a complicated sorbate system since it contains sodium (Na), Al, potassium (K), and magnesium (Mg), in addition to the intended HM ions, which are Cu, Zn, Fe, and Pb [16]. In these situations, the elimination of HM might be negatively affected by the selected resins' low desorption ability.

The sorptive efficacy of derivatives produced from oxygen (O), nitrogen (N), and sulfur (S) could potentially be worth considering in relation to HM extraction from such intricate sorbate systems. In accordance with the Hard and Soft Acids and Bases (HSAB) concept, weak acids react more quickly and form stronger bonds with weak bases when all other factors remain the same, while strong acids do the same with strong bases. For extremely intricate sorbate systems, such as real-time-based commercial discharges, the theory requires validation *via* further experiments [17]. Standard resins and modified chitosan-based substances can be analyzed in terms of their ability to effectively remove HMs from complex sorbate systems using the HSAB concept. However, the function of additional co-existing ions requires scientific validation, since they trigger relevant speciation and complex formation.

To date, numerous sorptive studies have been conducted to remove HM from effluent systems, utilizing a variety of sorbents. In several instances, researchers neglected to examine the cyclic reuse potential attributes of the wasted sorbents [18, 19]. Numerous researchers used complex and expensive eluents, including ethylenediaminetetraacetic acid disodium salt dihydrate (Na_2EDTA), ethylenediaminetetraacetic acid (EDTA), and EDTA-hydrochloric acid (HCl), for the purification of HMs from utilized sorbents. These eluents are more expensive, dangerous, and not long-lasting for large-scale situations that need HM removal by sorption and resin reuse [20]. It may be possible to use more straightforward, acidic, and basic eluents to successfully target the elimination of HMs by desorption. So, it is possible to target economically feasible multi-HM eradication. Recent developments in research also support the need for studies using simpler eluents. Nonetheless, certain studies focused on wastewater systems characterized by reduced chemical complexity. Consequently, investigations using basic eluents (sodium hydroxide (NaOH), HCl, nitric acid (HNO_3), sodium chloride (NaCl), potassium hydroxide (KOH), etc.) have been limited, particularly in relation to more intricate sewage systems [21].

For the purpose of removing HM from both acidic and water-based solutions, many sorbents have been documented. This has been clearly shown in the appropriate prior-art sub-sections. Nonetheless, there have been no contributions to examine the impact of chitosan's molecular weight on sorptive and desorptive effectiveness and resin reuse potential [22, 23]. It is also important to look at the proportions of the insertion and crosslinking agents that were employed. The previous art did not take this into account. These guidelines will help us learn more about how they affect the creation of cost-effective and environmentally friendly resins and the enhancement of processes.

The pertinent prior-art parts elucidated research aimed at the removal of HMs. But there has been no feedback regarding how much it costs. Consequently, it is essential to evaluate the cost-effectiveness of diverse sorbents for the extraction of HMs from complicated sorbate systems. Consequently, it is essential to engage in thorough introspection about the most effective resin identification approaches and to establish standards grounded in competition to elevate the standard of research in resin manufacturing science and the resin implementation sector. To sum up, despite their impressive capabilities, there are still some issues that need to be addressed before chitosan-based sorbents may be used in the real world to remove metallic ions. (i) Specific types of sorbents need to be more reliable. The sorption capability of many sorbents is limited by their tendency to degrade under particular harsh conditions, including very acidic solutions. (ii) The materials used to make sorbents may be quite expensive and even hazardous to the environment. Environmental concerns and increased fabrication costs are inevitable outcomes of this. (iii) Adsorbents with superior sorption configurations always run into big problems. (iv) The absorption potential of chitosan has been repeatedly evaluated using rather high feed levels in the reported research. The number of metallic ions in the contaminated waterways is consistently very low, less than 1.0 ppm. Consequently, the laboratory inquiry and the practical applications diverge. Moreover, while it seems that some investigators have explored the influence of supplementary factors on metal elimination, we discovered that most studies have inadequately addressed the issue of specificity.

4.3 Comparative Efficacy of Chitosan-Derived Sorbents in Wastewater Treatment

Chitosan-derived sorbents have shown great ability to remove organic particles, pigments, HMs, and phenolic chemicals [24]. Many studies have shown that chitosan can adsorb various kinds of metals, including HM cations (Pb, Cu, Hg, Cd), HM oxyanions (As and Cr), and precious ions (platinum (Pt), gold (Au), and palladium (Pd)) [25]. It was also clear that chitosan could be regenerated and used again as a sorbent. Due to this property, it has been shown to be a good way to remove Zn and other metal ions from wastewater [26]. Grafting, interlinking, and modification may make chitosan's mechanical characteristics better and more stable. Because the reactions are so simple, it is a great part for use in hybrid materials [24]. The successful elimination of Cr (67.66 mg/g) using modified chitosan, in contrast to ethylenediamine-functionalized iron oxide (Fe_3O_4) (61.35 mg/g) and poly-(4-vinyl pyridine) modified activated carbon (C) (53.7 mg/g), has shown its applicability in sorption processes. Many studies have demonstrated that evolved chitosan may be used to desalinate water [27]. The demand for low-cost, highly effective chitosan-derived solutions means that there is a lot of promise for water treatment in the near future. Table 4.2 shows a list of several isotherms and kinetic approaches that have been employed to learn more about how contaminants stick to amended chitosan. Table 4.3 shows the best values for several adsorption properties.

4.4 Challenges and Prospects

4.4.1 Strategies for the Real World Rather Than in a Laboratory Setting

Even though chitosan-derived resins have shown great promise, much of the research has been carried out in research laboratories, and they haven't yet been widely used in wastewater treatment in industry. Since it has a lot of different and complicated parts, industrial effluent is a big problem. This shows that we need biosorbents that are personalized and produced in a way that is good for the environment, and can be used to remove a wide range of contaminants [54]. Future research should concentrate on the development of chitosan-derived hybrid materials for the specific sorption of contaminants from intricate, practical substitute sorbate systems. In order to accomplish this, it is essential to focus on environmentally friendly fabrication techniques, enhance the durability of hybrid materials in harsh acidic and alkaline conditions, and establish secure disposal procedures.

4.4.2 Industrialization Encounters

There are several problems that need to be solved before chitosan-derived sorbents can be used on a broad scale and widely adopted. The primary sources of initial components are the skeletal systems of crustacean species and the cell structures of fungi, which are rather pricey. Thus, it is necessary to spend a lot of money and time to make sure that every part of making chitosan is pure. The expenses of extracting and purifying chitosan are also

Table 4.2 A brief overview of real-life mathematical models for kinetics, isotherms, and thermodynamics studies.

S. no.	Model		Equations	Significance	References
1	**Isotherms**	Langmuir isotherm	$\frac{C_e}{q_e} = \frac{1}{q_{max}} c_e + \frac{1}{K_{Lq_{max}}}$	Commonly utilized for the adsorption of solutes from aqueous solutions. It is assumed that all binding sites are uniformly distributed on the adsorbent surface and exhibit the same affinity for the adsorbate.	[28]
		Freundlich isotherm	$\ln q_e = \ln K_F + \frac{1}{n} \ln c_e$	It indicates that adsorption is heterogeneous. Multilayer adsorption. Non-uniform active sites and energy redistribution. The n variable indicates the resistance to adsorption. Values that are above 1.0 suggest that adsorption is not that complex. If n is below 1.0, the process is described as physical adsorption, while values above one indicate that it is chemical adsorption	

(Continued)

Table 4.2 A brief overview of real-life mathematical models for kinetics, isotherms, and thermodynamics studies. (*Continued*)

S. no.	Model		Equations	Significance	References
2	**Kinetics**	Pseudo first order	$\ln(q_e - q_t) = \ln q_e - \frac{k_1}{2.303}t$	It is the assumption that the rate of adsorption is directly related to the number of unoccupied sites. The model depicts physical adsorption at the surface of the adsorbent.	[29]
		Pseudo-second order	$\frac{t}{q_t} = \frac{1}{K_2 q_e^2} + \frac{1}{q_e}t$	The assumption is that the rate is directly proportional to both the number of vacant active sites and the concentration of the adsorbate.	[28]

(*Continued*)

Table 4.2 A brief overview of real-life mathematical models for kinetics, isotherms, and thermodynamics studies. (*Continued*)

S. no.	Model		Equations	Significance	References
3	Thermodynamics	Thermodynamics	$k_c = \frac{c_a}{c_e}$ $\Delta G = -RT\ln k_c$ $\Delta G = \Delta H - T\Delta S$	With regards to adsorption, enthalpy helps to reveal operational limits and its endothermic or exothermic nature. An enthalpy that is defined as positive signifies endothermic adsorption, while negative values define exothermic adsorption. The feasibility of adsorption is captured through Gibbs free energy; if it is negative, then adsorption transpires without any assistance from external forces.	[30]

Table 4.3 Various chitosan-derived adsorbents used for the adsorption of hazardous metal ions.

S. no.	Adsorbent	Metals	Dose (g/L)	pH	Concentration (mg/L)	Time (min)	Sorption capability (mg/g)	Desorption %	References
1	Chitosan-alginate@ Fe/Mn mixed oxide	Chromium (Cr)	1.0	4.0	25.0–1000.0	120	455.9	70 (NaOH) (5 Cycles)	[31]
2	CS/PAT/MWCNT@ MgAlLDHs	Cadmium (Cd)	3.0	8.0	100.0–500.0	30	1106.19	83 (NaOH) (3 Cycles)	[32]
3	β-cyclodextrin/ grafted-chitosan	Nickel (Ni) and Cobalt (Co)	0.5	5.0	100.0	120	154.0, 160.0	40 (0.01 M HCl) (5 Cycles)	[33]
4	Chitosan-vermiculite composite	Copper (Cu) and Cadmium (Cd)	10.0	8.0	250.0	360	116.22, 147.64	-	[34]

(Continued)

Table 4.3 Various chitosan-derived adsorbents used for the adsorption of hazardous metal ions. (*Continued*)

S. no.	Adsorbent	Metals	Dose (g/L)	pH	Concentration (mg/L)	Time (min)	Sorption capability (mg/g)	Desorption %	References
5	Citric acid/medium MW CS	Zinc (Zn), Lead (Pb), and Iron (Fe)	1.4	3.64	194.9–584.7 2.65–7.95 104.8–314.4	420	243.90 3.93 144.93	57.76 49.83 58.35 (2.0 M HCl) (3 Cycles)	[35]
6	Carboxymethyl/ medium MW CS						238.10 4.78 71.24	64.51 53.73 61.82 (2 M HNO_3) (3 Cycles)	[36]
7	Magnetic chitosan-based composite microparticles	Arsenic (As^{+3}, As^{+5})	2.0	7.0	0.5–7.0	-	33.68 34.61	90 (0.1 N NaOH) (10 Cycles)	[37]

(*Continued*)

Table 4.3 Various chitosan-derived adsorbents used for the adsorption of hazardous metal ions. (*Continued*)

S. no.	Adsorbent	Metals	Dose (g/L)	pH	Concentration (mg/L)	Time (min)	Sorption capability (mg/g)	Desorption %	References
8	Polyvinyl alcohol/ high MW CS	Zinc (Zn), Lead (Pb), and Iron (Fe)	1.4	3.64	194.9–584.7 2.65–7.95 104.8–314.4	360	222.21 4.02 135.14	51.89 53.70 59.94 (0.1 M HCl) (3 Cycles)	[38]
		Copper (Cu), Lead (Pb), and Iron (Fe)	1.6	3.82	187.7–563.1 5.2–15.6 61.9–185.5	240	243.9 6.65 87.72	44.3 31.9 53.7 (0.1 M HCl) (3 Cycles)	[39]
9	Carboxymethyl/ medium MW CS		1.2			540	344.83 9.59 90.09	61.52 48.11 44.44 (2 M HNO_3) (3 cycles)	[40]

(*Continued*)

Table 4.3 Various chitosan-derived adsorbents used for the adsorption of hazardous metal ions. (*Continued*)

S. no.	Adsorbent	Metals	Dose (g/L)	pH	Concentration (mg/L)	Time (min)	Sorption capability (mg/g)	Desorption %	References
10	Chitosan-conjugated iodate	Copper (Cu)	0.1	5.0	10.0–125.0	70	95.2	-	[41]
11	Iron manganese oxide gel beads (CS-FMO)	Arsenic (As^{+5}, As^{+3})	1.0		0.0–100.0	1200	26.65 35.96	60 95 (NaClO) (10 Cycles)	[42]
12	Poly (4-styrenesulfonic acid-co-maleic acid) (PSSMA)-functionalized-magnetic chitosan	Copper (Cu), and Nickel (Ni)	0.6	5.0 6.0	500.0–1200.0	30 40	141.70 113.07	93 93 (5 Cycles)	[43]
13	High-performance service (HPS)-aerogel	Copper (Cu)	0.2	5.0	50.0–200.0	1440	50.21	-	[44]

(*Continued*)

Table 4.3 Various chitosan-derived adsorbents used for the adsorption of hazardous metal ions. (*Continued*)

S. no.	Adsorbent	Metals	Dose (g/L)	pH	Concentration (mg/L)	Time (min)	Sorption capability (mg/g)	Desorption %	References
14	Modified chitosan adsorbents: powder (C-emimAc) bead (CB-emimAc) sponge (CS-emimAc)	Cadmium (Cd)	0.25	8.0	10.0–50.0	60	72.99 84.75 55.25	90.9 94.7 93.5 (0.1 M HNO_3)	[45]
15	Glycidyl chloride-modified chitosan-Schiff base (CS/24Cl/ECH)	Lead (Pb)	0.3	5.0	-		170.0	-	[46]
16	Chitosan, Graphene Oxide, Reduced Iron Oxide (CSGO-R@IO) hydrogel	Copper (Cu)	0.4	6.0	2.0–200.0	180	119.5	-	[47]
17	Chitosan-iron-oxide-azolla Pinnata	Chromium (Cr)	2.0	4.0	50.0–500.0	15	294.12	85 (HCl) (5 Cycles)	[48]
18	Chitosan-modified diatomite	Arsenic (As^{+5})	3.0	5.0	0.5–20.0	60	5.10	-	[49]

(*Continued*)

Table 4.3 Various chitosan-derived adsorbents used for the adsorption of hazardous metal ions. (*Continued*)

S. no.	Adsorbent	Metals	Dose (g/L)	pH	Concentration (mg/L)	Time (min)	Sorption capability (mg/g)	Desorption %	References
19	Green-based Zero-Valent Iron/Porous Cellulose/Chitosan Composite Spheres (ZVI/PCSs)	Cadmium (Cd)	-	7.0	10.0–500.0	240	110.3	50 (0.01 M HCl) (3 Cycles)	[50]
20	Shellfish waste-derived mesoporous chitosan	Arsenic (As^{+5})	20.0	5.0	4.0–30.0	90	12.32	70 (0.5 N NaOH) (6 Cycles)	[51]
21	Xanthate-modified magnetic iron oxide silica (Fe_3O_4@SiO_2)-based polyvinyl alcohol/chitosan	Cadmium (Cd), Copper (Cu), and Lead (Pb)	1.0	5.5	-	120	307.0 100.0 67.0	-	[52]
22	Glycidyl chloride cross-linked chitosan@ magnetic iron oxide (Fe_3O_4)/ activated carbon	Mercury (Hg)	0.4	7.0	10.0–80.0	20	166.6	84.8 (0.5 M HCl) (5 Cycles)	[53]

quite significant. To make composites easier to make and more reliable, we need to get over the technological problems that still remain. This necessitates thorough research tactics that foster effective inventiveness for the long-term and cost-efficient advancement of chitosan products. It is also very difficult to determine the inherent properties of chitosan when synthesis processes are adding new functions to it. This will be meticulously addressed by optimizing manufacturing methods to get the necessary attributes. Case studies on these types of research deserve to be released and made accessible in open-access journals and other broad sources.

4.4.3 Innovations in Regeneration Strategies

There is an increasing interest in the repurposing of chitosan-derived hybrid materials. Nevertheless, a substantial constraint exists within information frameworks linked to scientific routes, influencing variables and costs related to rejuvenation. In order to resolve this knowledge deficit, additional research must prioritize the exhaustive examination of the regeneration procedures for chitosan-derived hybrid materials. This involves examining the microstructural alterations that transpire throughout several revival methods, identifying the essential factors that affect restoration performance, and optimizing conditions for maximum revival effectiveness while preserving the product's inherent attributes. To make the regeneration of chitosan compounds more financially viable, research will look into affordable chemicals, better ways to rejuvenate them, and lower electricity requirements. A variety of rejuvenation techniques, including thermal, chemical, and biological approaches, will be evaluated for their effectiveness and longevity. Innovative approaches, such as photocatalytic and microbiological rejuvenation, may provide intriguing replacements to conventional rejuvenation procedures.

4.4.4 Machine Learning (ML) and Artificial Intelligence (AI) Algorithms

The complex process of making chitosan-derived hybrid materials with numerous limitations requires careful awareness of the complicated interactions between many elements. Traditional experimental methodologies are laborious, costly, and can lead to inefficient data handling. To address these issues, the creation of accurate computational algorithms is essential. Subsequent research should use numerical modeling, artificial intelligence (AI), and machine learning (ML) expertise to provide highly precise and environmentally friendly circumstances for the optimum fabrication and deployment of chitosan-derived compounds. To find previously unknown connections between lab factors and efficacy, algorithms using ML may have to go through huge data sets. The algorithms reduce the number of required trials and associated costs by enabling the validation of the most advantageous ideal parameters and their associated values. By combining ML and big data analysis, investigators may make development and research more efficient, speed up the optimization procedure, and finally create superior chitosan-derived hybrids *via* quick progress in environmentally friendly production and tailored purposes.

4.4.5 Techno-Economic Considerations

As competition in the global market builds up, it is important to figure out if the chitosan-derived materials can make money. When doing this kind of study, we need to think about a number of important factors, including the cost of the basic components, the cost of production and finalization, the efficacy of the materials, their possible applications, and their ability to be recycled. Chitosan is an organic material that is both environmentally friendly and affordable. Most of it comes from coastal waste, including the shells of prawns and other crustaceans [55, 56]. The fisheries and aquaculture industries need to keep growing so that these basic requirements are always available at affordable prices. However, making chitosan production methods more efficient would have a direct effect on the costs of making and processing chitosan. New methods, including microwave-based or microbiological ones, are expected to lower the cost of making chitosan compared to the conventional methods of extraction. Manufacturing facilities and technological advancements are anticipated to markedly decrease the production costs of various high-performance chitosan materials.

4.4.6 Greener Synthesis

Chitosan-derived synthesized hybrid materials have garnered increased interest in the past few decades owing to their potential in health care, water purification, and food packaging industries. But the finalization procedures have an effect on the surroundings that raises concerns about environmental sustainability. In this regard, it is important to emphasize the environmental benefits of chitosan-derived material manufacturing in order to create more sustainable procedures. Utilizing chitosan as the basis material for manufacturing helps the environment and makes us less reliant on assets that can't be replaced. Alternative synthesis procedures employ solvents that are better for the ecosystem, such as water or ionic liquids. Green manufacturing techniques function at low temperatures and pressures, require minimal power, and make minimal waste. They frequently devise catalyst-free methods or employ environmentally favorable catalysts, such as microbes or organic substances. This lessens the likelihood of catalyst degradation and lessens the amount of damage done to the ecosystem. Green approaches to synthesis also try to make as little waste as possible. This may be accomplished by maximizing reaction circumstances, employing optimal quantities of reacting agents, utilizing effective elimination techniques, and creating environmentally friendly chitosan-derived materials that readily decompose in the surroundings.

4.5 Recommendations

This chapter provides an extensive overview of the present state of ecological restoration, connecting basic scientific ideas with the newest technical applications. The chapter does an excellent job of bringing together the many ways that physicochemical and biological strategies can potentially be employed for clearing up polluted locations. It provides the aspect in a way that is straightforward enough for beginners to understand and detailed enough to be beneficial for experts. The most significant aspect of this chapter is that it takes a broad and progressive perspective. It goes deeper than the technical details of cleanup to look at the larger context of ecological and commercial viability. It is an important tool for helping people identify the right procedure for a site by carefully looking at the kind of contamination, the soil's properties, and the surrounding circumstances. The chapter emphasizes that successful restoration is not merely a scientific issue but a complicated, social challenge by stressing the necessity for multidisciplinary cooperation amongst geological experts, chemical scientists, technologists, and lawmakers.

4.6 Conclusion

This chapter explores the use of chitosan derivatives as a versatile and effective solution for removing toxic metal ions from water. It explains the mathematical models used to interpret experimental data, identifies key research gaps, and offers recommendations for future study. Conventional methods often fall short in eliminating these pollutants, making chitosan-based adsorption a promising alternative. While significant progress has been made in developing new chitosan derivatives, more research is needed to create cost-effective ways to regenerate the adsorbents and to understand the interactions between chitosan and other grafting compounds. A key research area is the creation of new chitosan derivatives with specific, tailored properties, prioritizing sustainability, security, and effectiveness. Understanding these interactions is crucial to fully realizing the potential of chitosan derivatives, which have numerous applications in medicine, tissue engineering, and biosensors.

List of Abbreviations

AI	Artificial Intelligence
Al	Aluminum
As	Arsenic
Au	Gold
C	Carbon
Cd	Cadmium
Co	Cobalt
Cr	Chromium
CSGO-R@IO	Chitosan, Graphene Oxide, Reduced Iron Oxide
Cu	Copper
EDTA	Ethylenediaminetetraacetic acid
Fe	Iron
Fe_3O_4	Iron-oxide
HCl	Hydrochloric acid
Hg	Mercury
HMs	Heavy metals
HNO_3	Nitric acid
HPS	High-performance service
HSAB	Hard and Soft Acids and Bases
K	Potassium
KOH	Potassium hydroxide
Mg	Magnesium
ML	Machine Learning
N	Nitrogen
Na	Sodium
Na_2EDTA	Ethylenediaminetetraacetic acid disodium salt dihydrate
NaCl	Sodium chloride
NaOH	Sodium hydroxide

Ni	Nickel
NMs	Nanomaterials
O	Oxygen
Pb	Lead
Pd	Palladium
PSSMA	Poly (4-styrenesulfonic acid-co-maleic acid)
Pt	Platinum
S	Sulfur
SiO_2	Silica
Zn:	Zinc
ZVI/PCSs	Zero-Valent Iron/Porous Cellulose/Chitosan Composite Spheres

AI disclosure statement: The author acknowledges the use of AI-based tools for minor language editing or grammatical correction. However, all intellectual contributions, analyses, and conclusions presented in this chapter are entirely the author's own.

References

1. Patel, P.K. and Uppaluri, R.V., Adsorption of emerging pollutants utilizing chitosan derivatives: Recent advances and future perspective. *Int. J. Biol. Macromol.*, 299, 140203, 2025.
2. Bhardwaj, L.K., Kumar, D., Kumar, A., Phytoremediation potential of Ocimum sanctum: A sustainable approach for remediation of heavy metals, in: *Phytoremediation Potential of Medicinal and Aromatic Plants*, pp. 46–57, CRC Press, Boca Raton, 2023.
3. Patel, P.K., Singh, H.K., Singh, A., Waste Water Remediation and Nanoadsorbents, in: *Nanomaterials in Wastewater Research*, pp. 235–257, Springer, Singapore, 2025.
4. Kumar, A., Bauddh, K., Srivastava, S. eds., *Phytoremediation Potential of Medicinal and Aromatic Plants: A Bioeconomical Approach*, CRC Press, Boca Raton, 2023.
5. Patel, P.K., Pandey, L.M., Uppaluri, R.V., Adsorption behavior and techno-economic analysis of citric acid-modified chitosan for heavy metal removal from wastewater. *Chem. Eng. Sci.*, 121868, 2025.
6. Mohan, B., Kamboj, A., Singh, K., Singh, G., Pombeiro, A.J., Ren, P., Metal-organic frameworks (MOFs) materials for pesticides, heavy metals, and drugs removal: Environmental safety. *Sep. Purif. Technol.*, 310, 123175, 2023.
7. Kumar, V. and Patel, P.K., Assessment of present air quality in Lucknow city and its impact on human health. *J. Air Pollut. Health*, 2025.

8. Bhardwaj, L.K., Sharma, S., Jindal, T., Estimation of physico-chemical and heavy metals in the lakes of Grovnes & Broknes Peninsula, Larsemann Hill, East Antarctica. *Chem. Afr.*, 6, 5, 2677–2694, 2023.
9. Mandvi, Patel, P.K., Singh, H.K., Performance analysis of machine learning models for AQI prediction in Gorakhpur City: a critical study. *Environ. Monit. Assess.*, 196, 10, 924, 2024.
10. Ojha, A., Jaiswal, S., Thakur, P., Mishra, S.K., Bioremediation techniques for heavy metal and metalloid removal from polluted lands: a review. *Int. J. Environ. Sci. Technol.*, 20, 9, 10591–10612, 2023.
11. Patel, P.K., Nagireddi, S., Uppaluri, R.V., Pandey, L.M., Batch adsorption characteristics of Dowex Marathon MSA commercial resin for Au (III) removal from synthetic electroless plating solutions. *Mater. Today Proc.*, 68, 824–829, 2022.
12. Patel, P.K., Pandey, L.M., Uppaluri, R., Multi-metal adsorption and cyclic desorption characteristics of Zn+ 2 and Cu+ 2 constituting multi-component synthetic wastewater system using commercial resins, in: *North-East Research Conclave*, pp. 3–27, Springer Nature Singapore, Singapore, 2022.
13. Nagireddi, S., Patel, P.K., Golder, A.K., Uppaluri, R.V., Recovery of PdEDTA2− complex from electroless plating solutions using nitrogen-functionalized chitosan derivative. *Polym. Bull.*, 82, 11, 6153–6172, 2025.
14. Wang, J. and Chen, C., Chitosan-based biosorbents: modification and application for biosorption of heavy metals and radionuclides. *Bioresour. Technol.*, 160, 129–141, 2014.
15. Patel, P.K., Pandey, L.M., Uppaluri, R.V., Conceptual techno-economic analysis of chitosan derivative resin and commercial sorbents for multi-heavy metal removal. *Sep. Purif. Technol.*, 362, 131732, 2025.
16. Morcali, M.H., Zeytuncu, B., Baysal, A., Akman, S., Yucel, O., Adsorption of copper and zinc from sulfate media on a commercial sorbent. *J. Environ. Chem. Eng.*, 2, 3, 1655–1662, 2014.
17. Patel, P.K. and Uppaluri, R.V., Environmental sustainability through adsorption: A review of chitosan's potential in dye pollution remediation. *Sustainable Chem. Pharm.*, 46, 102096, 2025.
18. Zawierucha, I., Kozlowska, J., Kozlowski, C., Trochimczuk, A., Sorption of Pb (II), Cd (II) and Zn (II) performed with the use of carboxyphenylresorcinarene-impregnated Amberlite XAD-4 resin. *Desalin. Water Treat.*, 52, 1–3, 314–323, 2014.
19. Mnasri-Ghnimi, S. and Frini-Srasra, N.J.A.C.S., Removal of heavy metals from aqueous solutions by adsorption using single and mixed pillared clays. *Appl. Clay Sci.*, 179, 105151, 2019.
20. Ren, Y., Abbood, H.A., He, F., Peng, H., Huang, K., Magnetic EDTA-modified chitosan/SiO2/Fe3O4 adsorbent: preparation, characterization, and application in heavy metal adsorption. *Chem. Eng. J.*, 226, 300–311, 2013.

21. Nagireddi, S., Patel, P.K., Golder, A.K., Uppaluri, R.V., Evaluation of a triethylenetetramine-chitosan derivative for the removal of palladium ion from an intricate adsorbate system. *J. Mol. Struct.*, 1323, 140782, 2025.
22. Dev, V.V., Baburaj, G., Antony, S., Arun, V., Krishnan, K.A., Zwitterion-chitosan bed for the simultaneous immobilization of Zn (II), Cd (II), Pb (II) and Cu (II) from multi-metal aqueous systems. *J. Cleaner Prod.*, 255, 120309, 2020.
23. Benavente, M., Moreno, L., Martinez, J., Sorption of heavy metals from gold mining wastewater using chitosan. *J. Taiwan Inst. Chem. Eng.*, 42, 6, 976–988, 2011.
24. Nechita, P., Applications of chitosan in wastewater treatment, in: *Biological Activities and Application of Marine Polysaccharides*, vol. 1, no. 5395720, pp. 209–228, 2017.
25. Zubair, M., Arshad, M., Ullah, A., Chitosan-based materials for water and wastewater treatment, in: *Handbook of Chitin and Chitosan*, pp. 773–809, Elsevier, 2020.
26. Bhatt, P., Joshi, S., Bayram, G.M.U., Khati, P., Simsek, H., Developments and application of chitosan-based adsorbents for wastewater treatments. *Environ. Res.*, 226, 115530, 2023.
27. Qian, X., Li, N., Wang, Q., Ji, S., Chitosan/graphene oxide mixed matrix membrane with enhanced water permeability for high-salinity water desalination by pervaporation. *Desalination*, 438, 83–96, 2018.
28. Gautam, N. and Das, G., Adsorption of antiviral and antibiotics by recyclable biochar derived from Dillenia indica: Impact of pyrolysis temperature and duration. *Colloids. Surf., A*, 703, 135265, 2024.
29. Gautam, N., Deka, D., Das, G., Efficient adsorption of nitroaromatic compounds from reusable hierarchical porous biochar emanates from the Ananas comosus crown and Citrus limetta fibers. *Chem. Eng. Sci.*, 308, 121408, 2025.
30. Rajesh, Y., Srinu, N., Namrata, G., Ramgopal, U., Preparation, characterization and Pd (II) adsorption characteristics of chitosan–AC composites from electroless plating solutions. *Desalin. Water Treat.*, 84, 279–291, 2017.
31. Alqarni, L.S., Algethami, J.S., Billah, R.E.K., Alorabi, A.Q., Alnaam, Y.A., Algethami, F.K., Bahsis, L., Jawad, A.H., Wasilewska, M., López-Maldonado, E.A., A novel chitosan-alginate@ Fe/Mn mixed oxide nanocomposite for highly efficient removal of Cr (VI) from wastewater: Experiment and adsorption mechanism. *Int. J. Biol. Macromol.*, 263, 129989, 2024.
32. Rostami, M.S. and Khodaei, M.M., Preparation and characterization of CS/PAT/MWCNT@ MgAl-LDHs nanocomposite for $Cd2+$ removal and 4-nitrophenol reduction. *J. Environ. Health Sci. Eng.*, 22, 1, 179–195, 2024.
33. Norouzi, Z., Abdouss, H., Abdouss, M., Nickle (II) and cobalt (II) ion removal using electrospun β-cyclodextrin/grafted-chitosan nanofibers as efficient molecular filters. *Desalin. Water Treat.*, 317, 100215, 2024.
34. Salih, S.S., Shihab, M.A., Mohammed, H.N., Kadhom, M., Albayati, N., Ghosh, T.K., Chitosan-vermiculite composite adsorbent: Preparation,

characterization, and competitive adsorption of Cu (II) and Cd (II) ions. *J. Water Process Eng.*, 59, 105044, 2024.
35. Patel, P.K., Pandey, L.M., Uppaluri, R.V., Cyclic adsorption and desorption characteristics of citric acid-chitosan variant resins for Pb, Fe, and Zn removal from simulated mining and agricultural wastewater system. *J. Polym. Environ.*, 32, 11, 5750–5770, 2024.
36. Patel, P.K., Pandey, L.M., Uppaluri, R.V., Highly effective removal of multi-heavy metals from simulated industrial effluent through an adsorption process employing carboxymethyl-chitosan composites. *Environ. Res.*, 240, 117502, 2024.
37. Song, X., Nong, L., Zhang, Q., Liu, J., Zhang, S., Highly effective adsorption for both As (III) and As (V) in aqueous medium *via* magnetic chitosan-based composite microparticles: Low arsenic concentration, high column efficiency and adaptability. *J. Environ. Chem. Eng.*, 11, 5, 110874, 2023.
38. Patel, P.K., Pandey, L.M., Uppaluri, R.V., Adsorptive removal of Zn, Fe, and Pb from Zn dominant simulated industrial wastewater solution using polyvinyl alcohol grafted chitosan variant resins. *Chem. Eng. J.*, 459, 141563, 2023.
39. Patel, P.K., Pandey, L.M., Uppaluri, R.V., Cyclic desorption based efficacy of polyvinyl alcohol-chitosan variant resins for multi heavy-metal removal. *Int. J. Biol. Macromol.*, 242, 124812, 2023.
40. Patel, P.K., Pandey, L.M., Uppaluri, R.V., Synthesized carboxymethyl-chitosan variant composites for cyclic adsorption-desorption based removal of Fe, Pb, and Cu. *Chemosphere*, 340, 139780, 2023.
41. Pam, A.A., Elemile, O.O., Musa, D.E., Okere, M.C., Amos, O., Ameh, Y.A., Removal of Cu (II) *via* chitosan-conjugated iodate porous adsorbent: Kinetics, thermodynamics, and exploration of real wastewater sample. *Results Chem.*, 2023.
42. Liu, Y., Cai, L., Wang, X., Chen, Z., Yang, W., Efficient adsorption of arsenic in groundwater by hydrated iron oxide and ferromanganese oxide chitosan gel beads. *Sep. Purif. Technol.*, 315, 123692, 2023.
43. Wang, M. and You, X.Y., Efficient adsorption of antibiotics and heavy metals from aqueous solution by structural designed PSSMA-functionalized-chitosan magnetic composite. *Chem. Eng. J.*, 454, 140417, 2023.
44. Kuang, J., Cai, T., Dai, J., Yao, L., Liu, F., Liu, Y., Shu, J., Fan, J., Peng, H., 2023. High strength chitin/chitosan-based aerogel with 3D hierarchically macro-meso-microporous structure for high-efficiency adsorption of Cu (II) ions and Congo red. *Int. J. Biol. Macromol.*, 230, 123238, 2023.
45. Saheed, I.O., Yusof, E.N.M., Oh, W.D., Hanafiah, M.A.K.M., Suah, F.B.M., Fabrication of chitosan@ activated carbon composites in EmimAc for Cd (II) adsorption from aqueous solution: experimental, optimization and DFT study. *Int. J. Biol. Macromol.*, 242, 124798, 2023.
46. Xiao, L., Shan, H., Wu, Y., Chitosan cross-linked and grafted with epichlorohydrin and 2, 4-dichlorobenzaldehyde as an efficient adsorbent for removal

of Pb (II) ions from aqueous solution. *Int. J. Biol. Macromol.*, 247, 125503, 2023.
47. Ali, A., Khan, S., Garg, U., Luqman, M., Bhagwath, S.S., Azim, Y., Chitosan-based hydrogel system for efficient removal of Cu [II] and sustainable utilization of spent adsorbent as a catalyst for environmental applications. *Int. J. Biol. Macromol.*, 247, 125805, 2023.
48. Sarojini, G., Kannan, P., Rajamohan, N., Rajasimman, M., Bio-fabrication of porous magnetic Chitosan/Fe3O4 nanocomposite using Azolla pinnata for removal of chromium–Parametric effects, surface characterization and kinetics. *Environ. Res.*, 218, 114822, 2023.
49. Ma, L., Xu, H., Xie, Q., Chen, N., Yu, Q., Li, C., Mechanism of As (V) adsorption from aqueous solution by chitosan-modified diatomite adsorbent. *J. Dispersion Sci. Technol.*, 43, 10, 1512–1524, 2022.
50. Li, B., Li, M., Zhang, P., Pan, Y., Huang, Z., Xiao, H., Remediation of Cd (II) ions in aqueous and soil phases using novel porous cellulose/chitosan composite spheres loaded with zero-valent iron nanoparticles. *React. Funct. Polym.*, 173, 105210, 2022.
51. Billah, R.E.K., Islam, M.A., Lgaz, H., Lima, E.C., Abdellaoui, Y., Rakhila, Y., Goudali, O., Majdoubi, H., Alrashdi, A.A., Agunaou, M., Soufiane, A., Shellfish waste-derived mesoporous chitosan for impressive removal of arsenic (V) from aqueous solutions: A combined experimental and computational approach. *Arabian J. Chem.*, 15, 10, 104123, 2022.
52. Wang, S., Liu, Y., Yang, A., Zhu, Q., Sun, H., Sun, P., Yao, B., Zang, Y., Du, X., Dong, L., Xanthate-modified magnetic Fe3O4@ SiO2-based polyvinyl alcohol/chitosan composite material for efficient removal of heavy metal ions from water. *Polymers*, 14, 6, 1107, 2022.
53. Kaveh, R. and Bagherzadeh, M., Simultaneous removal of mercury ions and cationic and anionic dyes from aqueous solution using epichlorohydrin cross-linked chitosan@ magnetic Fe3O4/activated carbon nanocomposite as an adsorbent. *Diamond Relat. Mater.*, 124, 108923, 2022.
54. Patel, P.K., Singh, H.K., Singh, A., Nanomaterials in the Degradation of Toxic Organic Pollutants, in: *Nanomaterials in Environmental Remediation*, pp. 75–92, CRC Press, Boca Raton, 2025.
55. Patel, P.K. and Uppaluri, R.V., Chitosan and its functionalized derivatives for heavy metal ion elimination: A review of synthesis, mechanisms, and characterization studies. *Adv. Colloid Interface Sci.*, 103632, 2025.
56. Kumari, S., Chowdhry, J., Uppaluri, R.V., Patel, P.K., Chitosan-Based Materials: Bridging the Gap between Pollutant Removal and Sustainability. *Process Saf. Environ. Prot.*, 107671, 2025.

5

Chelation Strategies for Remediation of Heavy Metals from Municipal Solid Waste Compost: A Sustainable Approach

Vivek Raj[1*], Santosh Kumar Karn[2], Ashish Kumar[3], Laxmi Kant Bhardwaj[4], Arti Chamoli[2], Arun Kumar[5], Abhinav Chauhan[6] and Arun Lal Srivastav[7]

[1]Department of Biotechnology, Patna Women's College (Autonomous), Patna University, Patna, Bihar, India
[2]Department of Biochemistry & Biotechnology, Sardar Bhagwan Singh University, Dehradun, Uttarakhand, India
[3]Department of Chemistry, Patna Women's College (Autonomous), Patna University, Patna, Bihar, India
[4]Amity Institute of Environmental Toxicology, Safety and Management, Amity University, Noida, Uttar Pradesh, India
[5]Research Department, Mahavir Cancer Sansthan & Research Centre, Patna, Bihar, India
[6]Department of Botany, T. P. S College, Patliputra University, Patna, Bihar, India
[7]Department of Applied Sciences, School of Engineering and Technology, Chitkara University, Himachal Pradesh, India

Abstract

Due to the rapid increase in municipal solid waste (MSW) and electronic waste (e-waste) from household activities, waste has become a severe environmental pollution issue nowadays. Composting is one of the simple, reliable, and natural methods to manage the MSW, which involves the biodegradation of complex organic waste into simpler inorganic nutrients that could support agricultural yield and soil restoration. However, apart from its advantages, compost obtained from MSW is often polluted with toxic heavy metals (HMs) like cadmium (Cd), arsenic (As), lead (Pb), etc., which cause hazardous effects to the entire environment and human health. Hence, MSW-based compost should not be preferred for

Corresponding author: dr.vivekraj07bt@gmail.com; ORCID: https://orcid.org/0000-0002-4357-2506

Prangya Rath, Laxmi Kant Bhardwaj and Amit Kumar (eds.) Heavy Metal Remediation, (109–132)

farming purposes due to such properties. Therefore, there are various chemical leaching techniques for HM remediation, specifically chelating reagents, inorganic reagents, surfactants, and compound chemical reagents, which differ in their efficacy of remediation. This chapter has explored methods to eliminate HMs from MSW compost using chelate-assisted leaching, which uses chelating agents like ethylenediaminetetraacetic acid (EDTA) and glutamic acid diacetate (GLDA) to bind and extract HMs from the compost as an effective remediation technique towards the circular economy (CE). Numerous factors determine the success of this leaching process, including: quantity of the chelating agent employed, the pH, contact time, temperature, liquid-to-solid (L/S) ratio, and flow rate. These parameters are examined at batch and continuous processing conditions to maximize the HMs removal. This approach increases the quality and safety of compost, helps to protect the environment and health of the population. It aids in accomplishing Sustainable Development Goals (SDGs) by embracing sustainable waste management methods.

Keywords: Affecting factors, chelating agents, heavy metals, municipal solid waste, remediation, sustainable waste management

5.1 Introduction

Municipal solid waste (MSW) management poses a significant global obstacle, primarily due to urbanization and population growth. By the year 2050, it is projected that nearly 68 % of the world's people will live in urban regions, thereby intensifying the demand for effective waste management infrastructure [1]. This migration trend will pose more challenges for low- and middle-income nations, where urban expansion frequently exceeds waste management capabilities, leading to environmental, aesthetic, and public health issues. A prominent example is India, where urbanization is expected to reach 50 % of the total population by 2050, producing over 160,000 tons of MSW daily as of 2021 [2]. Globally, high-income nations will naturally produce waste with lesser organic content due to their reliance on packaged goods and industrial products, while lower-income countries will produce higher organic content waste because they will rely on food scraps, agricultural residues, and other biodegradable waste. This disparity poses tasks and prospects for composting and organic waste management. In low- and middle-income contexts, waste components analyses reveal that organic material can comprise up to 53 % of the waste stream in nations like India, highlighting composting as a viable strategy for MSW reduction [2].

However, many emerging nations experience deprivation the necessary organization for effective waste segregation and handling, resulting in an over-reliance on landfilling, which significantly contributes to greenhouse gas (GHG) emissions and leachate pollution [3]. India's "Policy on Promotion of City Compost" serves as a notable example of how developing nations can leverage composting for waste reduction, economic development, and soil improvement, although difficulties related to quality control and market acceptance persist. Cities in Europe and Japan have also managed to establish high-quality composting systems, with strong supporting regulations, ensuring that the end products of the compost are safe and marketable [4]. On the other hand, most developing nations use centralized plants that produce mixed-waste compost, highly contaminated with HMs, thus not encouraging its use in agriculture, and causing unsold piles [5]. Infrastructural development of composting in countries like India has opened up substantial avenues of employment, and currently, more than 3000 MSW compost plants are working and diverting waste mass away from landfills [6]. However, in many spheres, it is difficult to note high accomplishments even after financial investments and other incentives from the government. This realization focuses the requirement to have superior guidelines and market conformity to establish compost as a serious farming input rather than a byproduct waste. Improving on these attritions may allow developing countries to align more closely with international sustainability goals, encompassing the United Nations Sustainable Development Goals (UNSDGs) about accountable production and consumption, along with climate action [7]. Finally, even though the burden of waste management is somewhat different in emerging countries, international good practice, along with innovative technologies, offers useful insights. Development of sustainable MSW management, especially in composting, can play a crucial role in serving economic, social, and environmental goals, which are vital in federal and global sustainability efforts. This chapter discusses the MSW management in transitioning areas with emphasis on HM in composting, accompanied by chemical-assisted leaching involving chelating agents, including EDTA and GLDA, as an efficient, sustainable remediation method to close the circle to a CE, composting process, global examples, and market trends.

5.2 About Heavy Metal (HM) Contamination in Municipal Solid Waste (MSW) Compost

A detrimental impact on human health and the ecosystem has been observed with excessive exposure to HMs, which poses a serious challenge to the environment due to their toxicity, persistence in the environment, and the possibility of bioaccumulation. Some metals, including manganese (Mn), copper (Cu), molybdenum (Mo), iron (Fe), zinc (Zn), nickel (Ni), and chromium (Cr), are essential micronutrients that play vital roles in biological processes. However, excessive exposure can accumulate in critical organs, such as the liver, brain, and kidneys, leading to severe health problems [8]. In agricultural soils, HM pollution interferes with plant growth and also diminishes crop yields and threatens the food chain, particularly impacting vulnerable groups. Annually, large quantities of HMs, approximately one million tons of Ni along with five million tons of Pb, enter soils, primarily due to solid waste disposal, mining, and industrial activities [9]. The excessive use of pesticides and fertilizers worsens soil contamination. To solve this problem, several remediation methods of MSW compost are available, like surface capping, encapsulation, chemical stabilization, electrokinetic remediation, phytoremediation, photocatalytic degradation, etc. Among them, chemical leaching is the most effective and affordable method. In this technique, metals are dissolved using extraction solutions like chelators, surfactants, bases, salts, or acids. One must remember the costs involved and the threat of secondary pollution [2].

5.3 Environmental and Agricultural Needs

Use of wastes to manufacture fertilizer is important in ensuring that the agricultural sector is sustainable and the environment is properly managed [3]. Advanced technologies in MSW management can facilitate the effective bioconversion of waste into valuable resources and support recovery for agricultural systems. Organic waste, which is rich in organic materials and has minimal levels of contamination, is especially beneficial for this use. Through the processing of such garbage, we could greatly reduce the amount of waste generated in the world and at the same time reclaim nutrients, like potassium (K), carbon (C), phosphorus (P), and even nitrogen (N), which are vital in the maintenance of agricultural activities [10]. Composting can facilitate a CE by ensuring that the optimal value of resources is obtained by recycling organic wastes to produce nutrient-enriched products. Even though this process reduces waste and also relieves the strain on the environment, MSW compost usually faces the problem of contamination due to the processing of mixed waste [11]. Effective waste segregation remains difficult, attributed to excessive waste production and a shortage of social consciousness. Even though some technologies like air separators, trommels, and even magnets are utilized, they are often not enough to properly separate contaminants. As a result, MSW compost may contain toxic elements and HMs, exceeding acceptable limits by 70 to 90 %, making it unsellable [3].

Municipal solid waste (MSW) compost contains HMs that primarily originate from domestic hazardous waste. Despite constituting a small percentage, typically around 1 to 4 % of the municipal waste stream, domestic hazardous waste can heavily influence compost quality because of the concentration of hazardous materials [11]. Solid Waste Management (SWM) Rules, 2016, categorize domestic hazardous waste as materials including used batteries, compact fluorescent lamp (CFL) bulbs, thrown paint drums, and expired prescriptions [2]. The presence of polluted MSW compost could lead to the buildup of HMs in human tissues, resulting in muscle dysfunction, organ failure, anemia, and cancer [12]. Plants that absorb these metals experience stunted growth and reduced yields, endangering consumers, while pollutants progressively reduce soil fertility. This chapter seeks to improve our knowledge of MSW compost quality by analyzing current research on solutions for biodegradable treatment in a closed-loop CE, as illustrated in Figure 5.1. Compost devoid of HMs is a nutrient-rich substitute for artificial fertilizers, which are frequently expensive, hazardous, and non-biodegradable. The primary aim is to keep compost from turning back into waste and to establish its marketability.

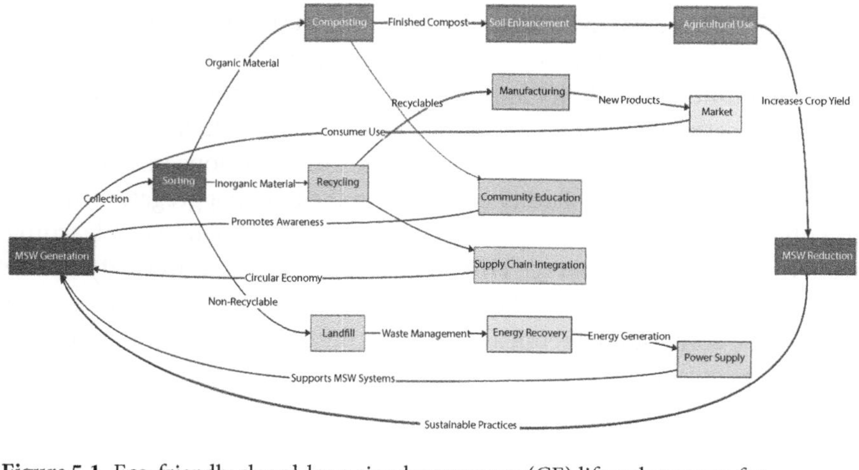

Figure 5.1 Eco-friendly closed-loop circular economy (CE) lifecycle process for municipal solid waste (MSW) compost.

5.4 Current Status of Heavy Metal (HM) Remediation in Municipal Solid Waste (MSW) Compost

Some of the techniques that could be utilized to remove HMs in MSW compost include chemical-assisted leaching, which has proved to be very cost-effective, besides being efficient. The methods also involve applying chemical solutions to leach HMs in compost, and this renders the compost usable in agriculture. It has also been demonstrated that biodegradable chelating agents, including GLDA, can eliminate up to 90.2 % of metals in polluted soils [13]. Besides their effectiveness in removing HMs, these materials also preserve the majority of soil nutrients; thus, they are environmentally friendly compared to the conventional chelators like EDTA. Moreover, the effectiveness of the leaching technique using chemicals has been cited to range between 70 to 95 %, depending on the chelating agent handled [2]. Besides being efficient, optimized chemical-assisted leaching processes have been observed to save greatly on operational cost when compared with other remediation methods, including electrokinetic remediation [14]. These results advocate the implementation of chemical leaching as both scalable and an economically feasible solution to the elimination of HMs in MSW compost, towards environmental sustainability.

5.4.1 Various Chemical-Assisted Leaching Techniques

The efficiency of the chemical-assisted leaching technique affects several factors, such as concentration, type, solid-liquid ratio, etc., of the leaching agent [15]. HMs exist in various states: carbonate-bound, organic-bound, water-soluble, iron-manganese oxide (Fe_2MnO_4) bound, and residual, each exhibiting different mobility and extractability [16]. In general, it is easier to remove carbonate-bound and water-soluble metals than those residual or bound to Fe_2MnO_4. The majority of metals reside as ions in acidic environments, but too much acidity can damage the macroaggregate structure and physicochemical characteristics of the matrix [11]. Choosing leaching agents that efficiently extract HMs while preserving the integrity of the matrix poses a significant challenge. These agents fall into four categories: surfactants, compound chemical reagents, chelating agents, and inorganic reagents, as outlined in Table 5.1.

(I) Inorganic Reagents

Inorganic acids, inorganic compounds, water, alkali, and redox reagents are examples of inorganic solutions frequently used. By destroying functional

Table 5.1 Classification of chemical agents by subtype, examples, mechanisms, and effects.

S. no.	Classification	Sub-type	Leaching mechanism	Advantages/disadvantages
1	Inorganic reagents	Salts, acids, water, alkali, and oxidizing or reducing reagents	Ion exchange, dissolution of metal compounds, complexation and valence change	**Advantages:** Removal efficacy is typically high; additionally, both acids and salts are broadly applicable and economical. **Disadvantage:** pH of the matrix is drastically changed by strong acids or alkalis, leading to structural damage and depletion of essential nutrients.
2	Surfactants	Chemical surfactants and biosurfactants	Modify surface properties, improve organic ligands' solubility in water, or use ion exchange to move pollutants from solid to liquid phase	**Advantages:** Chemical surfactants enhance the efficacy of heavy metal (HM) exclusion when mixed with additional washing agents. Likewise, biosurfactants, which are biodegradable and environmentally friendly, are also efficient at eliminating heavy metals (HMs). **Disadvantages:** Chemical surfactants by themselves are typically ineffective alongside not biodegradable. Biosurfactants, while generally not very effective at removal, are often expensive.
3	Compound chemical reagents	Mixed reagents, along with dissolved organic matter	Compound washing is dependent on the particular reagents used.	**Advantages:** It usually provides greater efficiency in the removal of heavy metals (HMs), greater environmental friendliness, and improved cost effectiveness. **Disadvantages:** Ideal results are not always achieved

(*Continued*)

Table 5.1 Classification of chemical agents by subtype, examples, mechanisms, and effects. (*Continued*)

S. no.	Classification	Sub-type	Leaching mechanism	Advantages/disadvantages
4	Chelating agents	Aminopolycarboxylic acids (APCs), and naturally low molecular organic acids	Forming stable chelates with heavy metal (HM) ions in matrix solution, along with desorbing heavy metals (HMs) from matrix particle surfaces	**Advantages**: Ethylenediaminetetraacetic acid (EDTA) works on a variety of heavy metals (HMs) over a broad pH range. However, ethylenediamine-N, N′-disuccinic acid (EDDS), iminodisuccinic acid (IDSA), and diethylenetriaminepentaacetic acid (DTPA) are biodegradable, while natural low molecular weight organic acids (NLMOAs) are both biodegradable along ecologically sustainable. **Disadvantages**: Ethylenediaminetetraacetic acid (EDTA), and diethylenetriaminepentaacetic acid (DTPA) are not very biodegradable, whereas nitrilotriacetic acid (NTA) is toxic and costly.

groups on the soil surface along with forming complexes with HMs by acidolysis, washing, and complexation or ion exchange, the soil washing mechanism with inorganic reagents primarily exchanges and desorbs HMs from soil to leachates [17].

(II) Surfactants

Surfactants are surface-active compounds that can significantly alter a solution's interfacial state. They are effective cleaning treatments for soils contaminated by HMs due to their ability to aid in solubilization, diffusion, and desorption of hydrophobic compounds [18]. Surfactants are classified into four categories: nonionic, anionic, cationic, and amphoteric. The primary organic amine derivatives that include N can dissociate cations in water and are known as cationic surfactants. Cationic surfactants are not often employed in soil washing because they hydrophobize the matrix surface and readily lead to secondary contamination in soil through adsorption. The most effective and popular surfactants are anionic ones, which are categorized as sulfonate and sulfate salts based on the hydrophilic group structure. Since nonionic surfactants' primary hydrophilic group is not a dissociated ether group in aqueous solution can be employed in a mixture with other surfactants because of their favorable compatibility. As a result, soil remediation frequently uses both anionic and nonionic surfactants. According to their natural and synthetic origins, surfactant molecules can be found in nature or as byproducts of laboratory synthesis. As a result, surfactants currently studied in soil remediation research and practice are typically divided into two categories: chemical surfactants and biosurfactants [18]. They remove HMs by sticking to the soil and forming micelles, which pull metal ions into the solution through an ionic process and complexation.

(III) Compound Chemical Reagents

Because of their different chemical characteristics, the coexisting metals make soil cleaning challenging, especially when combined with the requirement for soil protection. In this case, one reagent might not be enough to accomplish the remediation objective. Ethylenediaminetetraacetic acid disodium salt (Na_2EDTA), for example, is known to be the best chelating agent for removing cationic metals, not anionic metals [19]. The extraction of anionic metals from soils, sediments, and mine wastes is significantly higher for NLMOAs (including citric acid, oxalic acid, and tartaric acid) along with certain inorganic acids (including phosphoric acid and nitric acid) [17]. As a result, a new popular technique in recent years is the use of numerous solutions as compound chemical reagents to increase elimination efficacy. Compound chemical reagents are primarily separated into two categories: (a) mixed reagents, which are made up of multiple solutions, and (b) dissolved organic matter from different composts or wastes.

(IV) Chelation Assisted Leaching Technology

Chelating agents could combine with a number of metal ions to form stable, water-soluble compounds by chelation and desorption, generating favorable circumstances for soil leaching, leading to its remediation and recovery of HMs [17]. Many chelating reagents were studied for soil leaching, and these days, their functions are becoming increasingly important. "Aminopolycarboxylic acids" (APCs) and NLMOAs are two types of chelating reagents that are typically utilized in process recovery, research, and practice. APCs and NLMOAs are progressively employed for the environmental remediation of HMs from soil. APCs are artificial organic mixtures with multiple amine ($-NH_2$) and carboxylate (–COO–) groups, efficiently forming established metal facilities through coordination bonds, enhancing water solubility [20]. NLMOAs, also chelate metal ions, show more fragile stability than synthetic chelators [21]. APCs could be either non-biodegradable or biodegradable, while NLMOAs are inherently biodegradable [3]. Consequently, there is growing attention in investigating biodegradable, eco-friendly alternatives to traditional APCs for HM remediation [22]. Biodegradable substitutes like NTA and (S, S)-EDDS were proposed to substitute EDTA, although health concerns limit NTA's adoption [23]. Chelation technology in HM remediation represents a promising, sustainable technique for extracting HMs from dirty soils and MSW. Such a technique conserves resources and mitigates conservation risks connected with pollutants.

(a) Non-Biodegradable Chelating Agents

Chelating agents might be inorganic or organic complexes. Polyphosphates are the most well-known inorganic chelating agents; however, these are hydrolytically unpredictable at high temperatures and pH. Including organic chelating agents, APCs are commonly utilized chelants, which reported for a substantial share (39 %) in worldwide utilization of chelating agents in the year 2012 [2]. However, EDTA's preservation due to insignificant photodegradability and biodegradability can result in environmental accumulation and potential groundwater contamination, leading to health hazards to the biosphere [24]. These APCs, e.g., DTPA, NTA, EDTA, have a more durable combining capability than polyphosphates, along with excellent sequestering capability than hydrolytic acid-type chelating agents [25]. This type of chelation strategy creates established, water-soluble complexes by binding ligands to metal ions, which enhances solubility and facilitates elimination or retrieval.

(b) Biodegradable Chelating Agents

Such chelating agents are replacements for non-biodegradable alternatives like EDTA. Environmental addition of non-biodegradable chelating necessitates the development of sustainable alternatives. Confirming biodegradable chelators like GLDA, EDDS, IDSA, and NTA could alleviate the effect of HM-polluted solid matrices. Strength of metal chelation is affected by features like reaction kinetics, dissolved metal strengths, and specific soil characteristics. Key limitations influencing metal chelate strength include size and total number of rings developed, substituents on ligand rings, pH, metal category, and calcite strengths [26]. The stability constants (log K), indicating the intensity of metal-ligand interactions, provide insights into extraction efficacy. Bonding among metal ions and chelants varies based on oxidation states, coordination number, and the electronic features of chelants, encompassing a spectrum from electrostatic to covalent bonding. Although GLDA shows somewhat lower strength constants than EDTA, it remains valuable in HM complexation [27]. A study reviewed biodegradable chelators like GLDA and methylglycinediacetic acid (MGDA), finding that GLDA increased metal bioavailability by 70 % in contaminated soils while being 40 % cheaper than EDTA, making it a feasible option for global-scale remediation [28].

Researchers and coworkers pioneered a grouping of chelation and dechelation for Ni removal from spent catalysts, achieving a 98 % recovery rate alongside allowing for the reuse of the chelating agent [29]. An additional research team utilized a recyclable chelating agent (S, S)-EDDS for Ni removal, achieving 84 % revival under optimized circumstances [25]. Associated with conventional agents like (S, S)-EDDS, EDTA demonstrated effectiveness within a finer pH range, underscoring the advantages of biodegradable chelators for sustainable metal recovery. A study reported that a 0.1 M GLDA solution eliminated 65 % Cu and 78 % Pb, significantly outperforming conventional methods [30]. Their spectroscopic characterization confirmed that GLDA decreases HM binding with organic matter, enhancing extraction efficacy. The above follow-up research accomplished 85 % Zn and 92 % Cd removal utilizing a 0.15 M GLDA solution at 4.5 pH, exceeding EDTA in both efficacy and ecological compatibility [30]. A study demonstrated that biodegradable chelators at 0.2 M concentrations removed 80 to 90 % of Cu, Pb, and Zn from sludge compost [31]. Despite EDTA achieving slightly higher extraction efficacies, its environmental persistence raised worries, reinforcing the need for recyclable substitutes. A study found that adding 5 mM EDTA to turfgrass applications enhanced Pb and Cu uptake by 1.5 to 2.0 times compared to natural compost [30]. Integrating phytoremediation with a chelator-assisted technique can additionally enhance metal extraction while lessening chemical inputs. Another study emphasized the requirement for research on novel decomposable chelating agents, which can further advance environmentally friendly extraction methods [25]. Although promising, further study is essential to enhance the renewal and reclaim of biodegradable chelating agents, facilitating global-scale and constant metal removal systems. Based on Table 5.1, GLDA arises as the leading biodegradable chelating agent. It is a usual, plant-based compound that poses no health dangers, shows outstanding biodegradability (> 83 % within twenty-eight days), and additionally generates meaningfully lower carbon dioxide (CO_2) discharges during biodegradation associated with other chelating agents, making it an environmentally compliant alternative [20].

5.5 Major Parameters Influencing Efficiency of Chelation-Based Remediation

5.5.1 Process Parameters

5.5.1.1 Concentration of Chelating Agents

Both the type and concentration of chelators are essential in determining metal removal efficiency. HMs are divided into six different fractions using the European Community Bureau of Reference removal technique: exchangeable, water-soluble, acid-soluble, reducible, oxidizable, and residual [32]. Research indicates that leaching solutions primarily promote the dissolution, mobilization, and dispersion of inadequately bound metals, particularly those within acid-extractable and reducible portions. Subsequently, these mixtures combine with the metals to generate stable solubility complexes [33]. The concentration of the chelating agent is vital; higher concentrations give additional binding sites, thereby increasing extraction efficiency and complexation [34]. But rising EDTA concentrations beyond 0.10 to 0.20 M do not significantly improve the extraction of HMs, probably because of a reduction in available weakly bound metals [34]. At lower EDTA concentrations (ranging from 1.00 to 50.00 mM), elimination efficacy frequently suffers, potentially due to EDTA breakdown or competition with other metals, including calcium (Ca), Fe, aluminum (Al), and magnesium (Mg) [23, 34]. An increase in chelating agent concentration doesn't universally improve extraction across every HM; success is contingent upon specific metal-chelant interactions.

5.5.1.2 pH of the Chelating Solution

Variations in pH could modify the effectiveness of chelating agents by altering ion concentrations in solution, additionally influencing procedures including ion exchange with metal chelates, adsorption, and desorption [35]. Generally, lower pH levels enhance metal elimination by raising H^+ ion concentrations, which facilitates the dissociation of metals. But this relationship is complex. For example, a study found that utilizing polyacrylic acid (PAA) under highly acidic conditions to treat Cd and Pb-polluted soil diminished efficiency, as strong acidity weakened electrostatic forces surrounding a chelating agent, reducing its metal adsorption capacity [36]. Additionally, the redox potential (Eh) influences metal mobility, as Pb mobility is facilitated by lower Eh values and inhibited by higher Eh values.

5.5.1.3 *Liquid to Solid Ratio*

The liquid-to-soil (L/S) ratio is a critical factor for efficient leaching. Low L/S ratios limit interaction between the chelating agent and metals, thereby decreasing removal efficacy. Higher L/S ratios increase leachate volumes and energy requirements, necessitating careful resource management along with consideration of equipment usage and leachate management [37].

5.5.1.4 *Temperature*

Leaching temperature affects pollutant dissolution along with desorption [38]. Typically, higher temperatures improve the elimination of pollutants by enhancing solubility. But excessively greater temperatures might restrict the solubilization ability of certain leaching agents, potentially decreasing elimination efficiency. Notably, a temperature of 75 °C has demonstrated enhanced efficacy in specific cases [39]. Overall, temperature influences leaching outcomes less significantly than other scenarios, making large-scale temperature regulation less practical.

5.5.1.5 *Contact Time*

Leaching duration is a crucial element that governs the contact period between the chelating agent and HMs, immediately affecting the elimination rate. Metal detachment through a kinetically driven process of leaching notably influences the destiny of unstable metals [33]. Insufficient contact time may result in incomplete separation of metal, thereby increasing the toxicity of metals along with bioavailability because of weakly bound fractions [40]. Conversely, excessively prolonged leaching can lead to the dissolution of Fe and manganese oxides (MgO), causing phytotoxicity along with negative alterations in soil structure, including excessive mineral leaching along with pore clogging [41]. Leaching periods can range from a few hours to several days, influenced by soil features, kinds of metal, and their respective fractions [42].

5.5.1.6 *Agitation*

Agitation enhances the efficiency and effectiveness of batch leaching. It enhances the transmission of mass, confirms uniform mixing, decreases stagnant zones, and enhances kinetic performance [43]. Agitation also aids in temperature control and avoids sedimentation. While optimizing agitation achieves a balance between benefits and disadvantages, it might not always be necessary for satisfactory leaching results, based on the system and its aims.

5.5.1.7 *Flow Rate*

In column research, leaching solution flow rate substantially affects process effectiveness [44]. Increased flow rates lead to quicker breakthrough points, as shorter contact times result in faster breakthrough curve saturation. Higher flow rates enhance transfer of mass, improving contaminant-solution relations along bed height (mass-transfer zone). This rapid saturation of the mass-transfer zone enhances leaching efficacy; however, it decreases contaminant residence time [45]. Optimizing flow rate is essential for balancing mass transfer; moreover, contact time is required to achieve efficient solid matrix decontamination in continuous soil columns.

5.5.2 Operational Modes of Chelating

Several methods are utilized in systems for chemical leaching to treat polluted MSW compost, including batch, continuous moving bed, continuous fixed bed, continuous fluidized bed, and pulsed bed methods. Such approaches increase interaction between the leaching explanation and the constant matrix. Researchers commonly employ the batch method to analyze feasibility along with efficiency in leaching systems of chelating agents, as it permits controlled research along with optimization of experimental limitations.

5.5.2.1 *Batch Chelating*

In batch-mode chelating, input and output streams remain disconnected throughout the pollutant removal process (Figure 5.2). In this setup, the chelating solution and poisoned compost are introduced into the vessel and left untouched for a designated period. An orbital or agitator shaker might be employed to confirm mixing at a specific rotational speed [37]. Researchers periodically sample and analyze contaminant levels without any intermittent additions during the procedure. Because of its simplicity, minimal accessories, and low-cost requirements, this mode is favored by most researchers. It allows for a thorough comprehension and design parameters regulation, rendering it optimal for chemical leaching study [46].

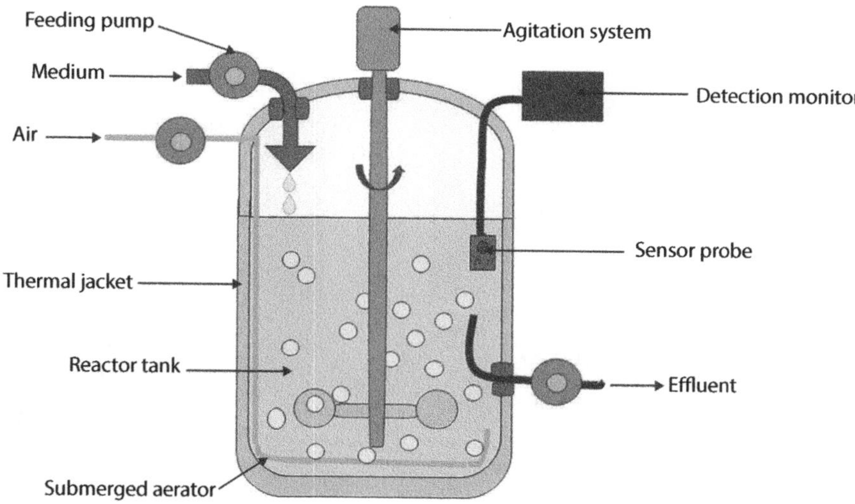

Figure 5.2 Batch-leaching reactor.

5.5.2.2 Continuous Column Mode of Operation

The fixed-bed column leaching procedure intended for constant flow operation is illustrated in Figure 5.3. In such a technique, infected objects are filled into a column of specific length, with strainers at the base preventing the leaching mixture. A positive displacement pump is connected to the column, including a peristaltic pump, which controls the flow rate [47]. As the result flows through the column, solutes travel and stay in the stationary solution unless equilibrium is reached. Such procedures continue until saturation occurs within the matrix's openings or functional spots [48]. Once saturation is achieved, solution transfer ceases. Process is influenced by diffusional resistances, needing adequate residence time. Such residence time, affected by flow rate, is essential for optimizing the leaching procedure [49]. Dynamic leaching facilitates scalability as per continuous data; however, it encounters difficulties like high pressure drops, non-ideal flow, and diffusional resistances, which limit practical implementation [47].

5.6 Proposed Future Research and Directions

Forthcoming research should focus on optimizing GLDA as a biodegradable chelating agent for HM remediation in MSW compost. Literature indicates that GLDA is an eco-friendly and efficient alternative to conventional

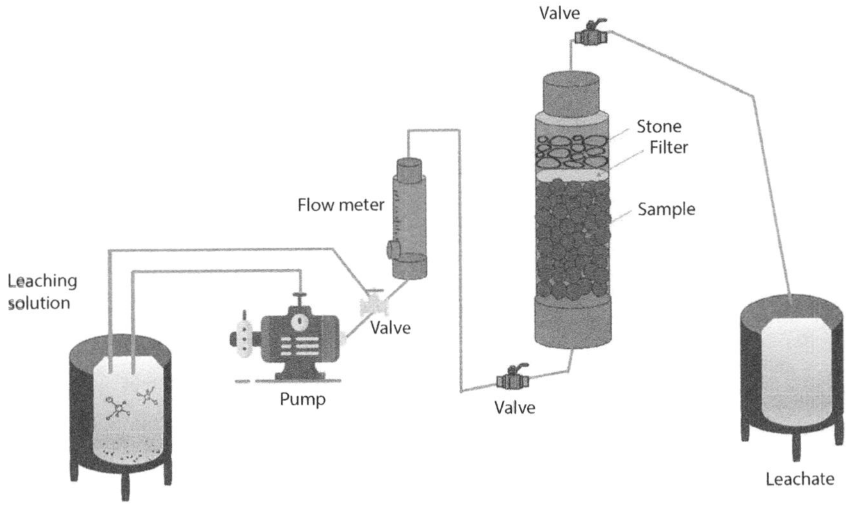

Figure 5.3 Continuous column leaching setup.

chelators like EDTA. The batch chelating technique is the best cost-effective approach, offering similar efficiency to column leaching at reduced operational costs. To enhance the chelating process, factors such as pH, chelating agent intensity, and contact time will be analyzed to maximize metal ion solubility and efficiency. Additional factors like temperature, agitation, L/S ratio, and particle size will be examined to improve mass transfer and response kinetics. This systematic approach will detect optimal environments for effective HM extraction and refine the process through optimization techniques.

A modified Tessier's sequential extraction will provide insights into the chelating mechanism by analyzing HM distribution in MSW compost before and after management. Categorization of metal fractions will help determine the efficiency of remediation in achieving safe levels for numerous land uses. Furthermore, pot experiments will assess plant growth and accumulation of HMs after using remediated compost. Use of a biodegradable chelating agent provides a feasible solution for removing hazardous metal ions and addressing climate change. This framework supports the UNSDGs, particularly in sustainable agriculture and environmental protection. The cost of chelating agents, energy, and labor will be used in the economic feasibility analysis, which will indicate the benefits of GLDA compared to EDTA in large-scale uses. There will be revenue potential from remediated compost, and this will be in accordance with the environmental regulations to increase marketability. Policy implications are crucial for the successful commercialization of this technology. Governments can promote sustainable waste management practices by offering subsidies and grants to waste treatment facilities that support the use of biodegradable chelating agents.

5.7 Recommendations

Municipal solid waste (MSW) composting enhances nutrient cycling by biodegrading urban organic waste into valuable agricultural inputs, linking agriculture and urban waste management. However, preserving best-quality, pollutant-free compost stays difficult due to ineffective waste separation. MSW conversion to compost is common globally, making it essential to separate waste by type. Compost can be applied directly to garden plants and field crops, but compost containing exchangeable metallic concentrations can lead to accumulation in plants and enter the food chain. Therefore, it is recommended that compost have zero exchangeable metallic concentration for optimal use, which must be checked before compost preparation and application.

5.8 Conclusion

The impact of home hazardous waste and HM contamination in MSW compost is the main topic of the bibliometric analysis, which highlights a general lack of knowledge in tackling this important issue. Poor quality frequently leads to compost being heaped up as waste instead of being a useful product. Given the potential of MSW compost, this chapter underlined the necessity of creating efficient remediation methods to eliminate HMs and improve compost quality. For managing HM contamination, chelation technology has emerged as a promising solution that balances social, environmental, and economic sustainability. The cost-effective retrieval and recycling of chelating agents, closed-loop operations with minimal releases or hazardous byproducts, and a safe soil claim that meets sustainability objectives, along with public reception, are all considered key features of chelator-based remediation. A shift toward sustainable solutions is shown by the recognition of biodegradable chelating agents as possible substitutes. There are still major obstacles in the areas of cost-effectiveness, reuse, and regeneration. Prolonged viability and safety of remediated compost as soil amendment needs more research, especially in light of HM remobilization and the effects it has on the environment, plants, microorganisms, and soil. Using biodegradable chelating agents like GLDA with enhanced pre- and post-leaching effectively removes HMs. Optimizing factors like GLDA concentration, pH, and retention time is crucial for removing HMs from MSW. Combining chemical methods with a bioremediation approach offers a comprehensive and sustainable solution. These strategies can develop a sustainable method to eliminate HMs, improve compost quality, enable safe use in agricultural fields, and align with the CE.

List of Abbreviations

Al	Aluminum
APCs	Aminopolycarboxylic acids
As	Arsenic
C	Carbon
Ca	Calcium
Cd	Cadmium
CE	Circular economy
CFL	Compact fluorescent lamp
CO_2	Carbon dioxide
Cr	Chromium
Cu	Copper
DTPA	Diethylenetriaminepentaacetic acid
EDDS	Ethylenediamine-N, N'-disuccinic acid
EDTA	Ethylenediaminetetraacetic acid
Eh	Redox potential
E-waste	Electronic waste
Fe	Iron
Fe_2MnO_4	Iron-manganese oxide
GHG	Greenhouse gas
GLDA	Glutamic acid diacetate
HMs	Heavy metals
IDSA	Iminodisuccinic acid
K	Potassium
L/S ratio	Liquid/solid ratio
Mg	Magnesium
MGDA	Methyl Glycine-N,N-Diacetic Acid
MgO	Manganese oxides
Mn	Manganese
MO	Molybdenum
MSW	Municipal solid waste

N	Nitrogen
Na_2EDTA	Ethylenediaminetetraacetic acid disodium salt
NI	Nickel
NLMOAs	Natural low molecular organic acids
NTA	Nitrilotriacetic acid
P	Phosphorus
PAA	Polyacrylic acid
Pb	Lead
SDGs	Sustainable Development Goals
SWM	Solid waste management
UNSDGs	United Nations Sustainable Development Goals
Zn	Zinc

AI disclosure statement: The author acknowledges the use of AI-based tools for minor language editing or grammatical correction. However, all intellectual contributions, analyses, and conclusions presented in this chapter are entirely the author's own.

References

1. Winkler, M.K. and Van Loosdrecht, M.C.M., Intensifying existing urban wastewater. *Science*, 375, 6579, 377–378, 2022.
2. Khare, S., Singhal, A., Rallapalli, S., Mishra, A., Bio-chelate assisted leaching for enhanced heavy metal remediation in municipal solid waste compost. *Sci. Rep.*, 14, 1, 14238, 2024.
3. IA, A. and Demehin, A.A., Assessment of Potential Health Hazards Associated with Bacterial Diversity in Municipal Solid Wastes of Okitipupa LGA in Ondo State, Nigeria. *Niger. J. Microbiol.*, 2024.
4. Zhang, Z., Chen, Z., Zhang, J., Liu, Y., Chen, L., Yang, M., Osman, A.I., Farghali, M., Liu, E., Hassan, D., Ihara, I., Municipal solid waste management challenges in developing regions: A comprehensive review and future perspectives for Asia and Africa. *Sci. Total Environ.*, 930, 172794, 2024.
5. EFSA Panel on Biological Hazards (EFSA BIOHAZ Panel), Koutsoumanis, K., Allende, A., Bolton, D.J., Bover-Cid, S., Chemaly, M., Davies, R., De Cesare, A., Herman, L.M., Hilbert, F., Lindqvist, R., Evaluation of alternative methods of tunnel composting (submitted by the European composting network). *EFSA J.*, 18, 8, e06226, 2020.
6. Barthod, J., Rumpel, C., Dignac, M.F., Composting with additives to improve organic amendments. A review. *Agron. Sustainable Dev.*, 38, 2, 17, 2018.

7. Hamdan, A., Panda, S., Jain, M.S., Raj, V., Mathew, S., Assessing municipal solid waste in Indian smart cities: A path towards Waste-to-Energy. *Heliyon*, 11, 6, 2025.
8. Korc, M. and Hauchman, F., Advancing environmental public health in Latin America and the Caribbean. *Rev. Panam. Salud Publica*, 45, e118, 2021.
9. Bhardwaj, L.K., Rath, P., Choudhury, M., Metzdorff, A., Jodder, P.K., Electronic wastes and contamination of water, in: *Electronic Waste*, pp. 48–57, CRC Press, Boca Raton, Florida, United States, 2025.
10. Moustakas, K. and Loizidou, M., Effective waste management with emphasis on circular economy. *Environ. Sci. Pollut. Res.*, 30, 4, 8540–8547, 2023.
11. Singh, R., Toolkit: legacy waste management and dumpsite remediation to support Swachh Bharat Mission 2.0, Centre for science and environment, New Delhi, pp. 1–72, 2022.
12. Hölzle, I., Somani, M., Ramana, G.V., Datta, M., Heavy metals in soil-like material from landfills–Resource or contaminants? *J. Cleaner Prod.*, 369, 133136, 2022.
13. Suanon, F., Sun, Q., Dimon, B., Mama, D., Yu, C.P., Heavy metal removal from sludge with organic chelators: comparative study of N, N-bis (carboxymethyl) glutamic acid and citric acid. *J. Environ. Manage.*, 166, 341–347, 2016.
14. Reddy, K.R. and Cameselle, C., *Electrochemical remediation technologies for polluted soils, sediments and groundwater*, John Wiley & Sons, Hoboken, New Jersey, USA, 2009.
15. Zhu, Z., Wang, J., Liu, X., Yuan, L., Liu, X., Deng, H., Comparative study on washing effects of different washing agents and conditions on heavy metal contaminated soil. *Surf. Interfaces*, 27, 101563, 2021.
16. Chen, X., Ren, Y., Li, C., Shang, Y., Ji, R., Yao, D., He, Y., Study on Factors Influencing the Migration of Heavy Metals from Soil to Vegetables in a Heavy Industry City. *Sustainability*, 16, 24, 11084, 2024.
17. Zhang, H., Xu, Y., Kanyerere, T., Wang, Y.S., Sun, M., Washing reagents for remediating heavy-metal-contaminated soil: a review. *Front. Earth Sci.*, 10, 901570, 2022.
18. Liu, J.H., Xue, J., Wei, X.H., Research progress of surfactant washing for remediation of heavy metal pollution in soil. *J. Soil Sci. (in Chinese)*, 50, 240–245, 2019.
19. Su, M., Qi, J., Xue, K., Yang, X., Liu, M., Ju, C., Peng, X., Wu, Y., Towards Efficient Extraction of Heavy Metals from Contaminated Soils: Soil Leaching Integrated with Flocculation Dewatering. *Water Air Soil Pollut.*, 235, 6, 398, 2024.
20. Begum, Z.A., Rahman, I.M., Tate, Y., Sawai, H., Maki, T., Hasegawa, H., Remediation of toxic metal contaminated soil by washing with biodegradable aminopolycarboxylate chelants. *Chemosphere*, 87, 10, 1161–1170, 2012.
21. Zhao, B., Tang, J., Lyu, H., Liu, F., Wang, L., Low molecular weight organic acids strengthen the electron transfer of natural FeS2/biochar composite for

Cr (VI) reduction: Experimental observations and governing mechanisms. *J. Environ. Chem. Eng.*, 10, 2, 107181, 2022.

22. Diarra, I., Kotra, K.K., Prasad, S., Assessment of biodegradable chelating agents in the phytoextraction of heavy metals from multi-metal contaminated soil. *Chemosphere*, 273, 128483, 2021.
23. Wang, G., Zhang, S., Xu, X., Zhong, Q., Zhang, C., Jia, Y., Li, T., Deng, O., Li, Y., Heavy metal removal by GLDA washing: optimization, redistribution, recycling, and changes in soil fertility. *Sci. Total Environ.*, 569, 557–568, 2016.
24. Shukla, M., Baksi, B., Mohanty, S.P., Mahanty, B., Mansi, A., Rene, E.R., Behera, S.K., Remediation of chromium contaminated soil by soil washing using EDTA and N-acetyl-L-cysteine as the chelating agents. *Prog. Org. Coat.*, 165, 106704, 2022.
25. Chauhan, G., Pant, K.K., Nigam, K.D.P., Chelation technology: a promising green approach for resource management and waste minimization. *Environ. Sci. Processes Impacts*, 17, 1, 12–40, 2015.
26. Chauhan, G., Jadhao, P.R., Pant, K.K., Nigam, K.D.P., Novel technologies and conventional processes for recovery of metals from waste electrical and electronic equipment: challenges & opportunities–a review. *J. Environ. Chem. Eng.*, 6, 1, 1288–1304, 2018.
27. Wei, Z., Chen, Y., Li, X., Rong, H., Huang, Z., Remediation of heavy metal contaminated farmland soil by biodegradable chelating agent GLDA. *Appl. Sci.*, 12, 18, 9277, 2022.
28. Yin, F., Li, J., Wang, Y., Yang, Z., Biodegradable chelating agents for enhancing phytoremediation: mechanisms, market feasibility, and future studies. *Ecotoxicol. Environ. Saf.*, 272, 116113, 2024.
29. Liang, X., Tang, J., Li, L., Wu, Y., Sun, Y., A review of metallurgical processes and purification techniques for recovering Mo, V, Ni, Co, Al from spent catalysts. *J. Cleaner Prod.*, 376, 134108, 2022.
30. Khare, S., Singhal, A., Rallapalli, S., Mishra, A., Bio-chelation for sustainable heavy metal remediation in municipal solid waste compost: A critical review of chelation technologies. *Environ. Sci. Pollut. Res.*, 1–25, 2025.
31. Hu, X., Huang, X., Zhao, H., Liu, F., Wang, L., Zhao, X., Gao, P., Li, X., Ji, P., Possibility of using modified fly ash and organic fertilizers for remediation of heavy-metal- contaminated soils. *J. Cleaner Prod.*, 284, 124713, 2021.
32. Chen, Y., Chen, Y., Li, Y., Wu, Y., Zeng, Z., Xu, R., Wang, S., Li, H., Zhang, J., Changes of heavy metal fractions during co-composting of agricultural waste and river sediment with inoculation of Phanerochaetechrysosporium. *J. Hazard. Mater.*, 378, 120757, 2019.
33. Zhang, W., Tsang, D.C., Chen, H., Huang, L., Remediation of an electroplating contaminated soil by EDTA flushing: chromium release and soil dissolution. *J. Soils Sediments*, 13, 354–363, 2013.
34. Xiao, R., Ali, A., Wang, P., Li, R., Tian, X., Zhang, Z., Comparison of the feasibility of different washing solutions for combined soil washing and

phytoremediation for the detoxification of cadmium (Cd) and zinc (Zn) in contaminated soil. *Chemosphere*, 230, 510–518, 2019.

35. Hasegawa, H., Al Mamun, M.A., Tsukagoshi, Y., Ishii, K., Sawai, H., Begum, Z.A., Asami, M.S., Maki, T., Rahman, I.M., Chelator-assisted washing for the extraction of lead, copper, and zinc from contaminated soils: A remediation approach. *Appl. Geochem.*, 109, 104397, 2019.
36. Feng, W., Zhang, S., Zhong, Q., Wang, G., Pan, X., Xu, X., Zhou, W., Li, T., Luo, L., Zhang, Y., Soil washing remediation of heavy metal from contaminated soil with EDTMP and PAA: Properties, optimization, and risk assessment. *J. Hazard. Mater.*, 381, 120997, 2020.
37. Chen, Y., Jiang, H., Li, Y., Liu, Y., Chen, Y., Chen, L., Luo, X., Tang, P., Yan, H., Zhao, M., Yuan, Y., A critical review on EDTA washing in soil remediation for potentially toxic elements (PTEs) pollutants. *Rev. Environ. Sci. Bio/Technol.*, 21, 2, 399–423, 2022.
38. Lemraski, E.G. and Sharafinia, S., Kinetics, equilibrium and thermodynamics studies of Pb2+ adsorption onto new activated carbon prepared from Persian mesquite grain. *J. Mol. Liq.*, 219, 482–492, 2016.
39. Meng, X., Meng, L., Gao, Y., Li, H., A comprehensive review on the spray cooling system employed to improve the summer thermal environment: application efficiency, impact factors, and performance improvement. *Build. Environ.*, 217, 109065, 2022.
40. Kördel, W., Bernhardt, C., Derz, K., Hund-Rinke, K., Harmsen, J., Peijnenburg, W., Comans, R., Terytze, K., Incorporating availability/bioavailability in risk assessment and decision making of polluted sites, using Germany as an example. *J. Hazard. Mater.*, 261, 854–862, 2013.
41. Beiyuan, J., Tsang, D.C., Bolan, N.S., Baek, K., Ok, Y.S., Li, X.D., Interactions of food waste compost with metals and metal-chelant complexes during soil remediation. *J. Cleaner Prod.*, 192, 199–206, 2018.
42. Tang, J., He, J., Liu, T., Xin, X., Removal of heavy metals with sequential sludge washing techniques using saponin: optimization conditions, kinetics, removal effectiveness, binding intensity, mobility and mechanism. *RSC Adv.*, 7, 53, 33385–33401, 2017.
43. Kyzas, G.Z., Favvas, E.P., Kostoglou, M., Mitropoulos, A.C., Effect of agitation on batch adsorption process facilitated by using nanobubbles. *Colloids Surf., A*, 607, 125440, 2020.
44. Elmaslar-Özbaş, E. and Balkaya, N., Extraction of heavy metals from compost using a mixture of Na2EDTA and Na2S2O5: column studies. *J. Soil Sci. Plant Nutr.*, 12, 3, 525–534, 2012.
45. Cánovas, C.R., Pérez-López, R., Macías, F., Chapron, S., Nieto, J.M., Pellet-Rostaing, S., Exploration of fertilizer industry wastes as potential source of critical raw materials. *J. Cleaner Prod.*, 143, 497–505, 2017.
46. Etim, E.U., Batch leaching of Pb-contaminated shooting range soil using citric acid- modified washing solution and electrochemical reduction. *Int. J. Environ. Sci. Technol.*, 16, 3013–3020, 2019.

47. Thirunavukkarasu, A., Nithya, R., Sivashankar, R., Continuous fixed-bed biosorption process: A review. *Chem. Eng. J. Adv.*, 8, 100188, 2021.
48. Xie, T., Li, X., Sun, H., Dan, Z., Characteristics and factors that influence heavy metal leaching from spent catalysts. *Environ. Sci. Pollut. Res.*, 29, 42, 63393–63406, 2022.
49. Feng, J., Yu, J., Huang, S., Wu, X., Zhou, F., Xiao, C., Xu, Y., Chi, R., Effect of potassium chloride on leaching process of residual ammonium from weathered crust elution- deposited rare earth ore tailings. *Miner. Eng.*, 163, 106800, 2021.

6

Sustainable Approaches to Heavy Metals Remediation

Prangya Rath[1], Laxmi Kant Bhardwaj[2*] and Amit Kumar[3]

[1]Amity Institute of Environmental Sciences, Amity University, Noida, Uttar Pradesh, India
[2]Amity Institute of Environmental Toxicology, Safety and Management, Amity University, Noida, Uttar Pradesh, India
[3]Molecular Biology and Biotechnology Division, CSIR-National Botanical Research Institute, Lucknow, Uttar Pradesh, India

Abstract

Cadmium (Cd), lead (Pb), mercury (Hg), arsenic (As), nickel (Ni), and chromium (Cr) are the chief heavy metals (HMs) that have triggered dangerous health and environmental concerns. These toxic pollutants can be accumulated in food chains. Ion exchange, electrochemical treatments, and chemical precipitation are the different conventional methods of remediation that have been reported to be high-energy-consuming and costly, and may lead to the production of secondary waste. Bioremediation is a technique that involves the use of microorganisms that are capable of bioaccumulating, bio-sorbing, and enzymatically transforming HMs into less toxic ones. In contrast, the phytoremediation technique is based on the usage of plants that take up and stabilize HMs in their tissues. Hence, it prevents the release of HMs into water and soil and their availability to organisms. In nano-biotechnology, the application of green, synthesized nanoparticles (NPs) is one of the ways to enhance metal adsorption and stability without causing any ecological impact. Advances in genetic engineering can positively influence the tolerance of a plant or microbe, as well as their uptake efficiency, and can also be a source of precision bioremediation. Sustainable HMs remediation methods are cost-effective and eco-friendly for removing HMs with the support of circular economy (CE) principles, which emphasize facets such as life cycle assessment (LCA) and the use of minimum resources. This study aims to describe the conventional as well

Corresponding author: bhardwaj.laxmikant@gmail.com; ORCID: https://orcid.org/0000-0001-7518-4199

Prangya Rath, Laxmi Kant Bhardwaj and Amit Kumar (eds.) Heavy Metal Remediation, (133–170)

as sustainable and green approaches of HM remediation. This study also describes the challenges, future perspectives, and provides some recommendations.

Keywords: Bioremediation, circular economy, genetic engineering, heavy metals, nanotechnology, phytoremediation, sustainability

6.1 Introduction

The pollution of heavy metals (HMs) can be harmful to the health of humans and the ecosystem [1]. These metals, such as Pb, Cd, Hg, As, Cr, and Ni, are stored in the environment for an extended time period as they are not decomposable. When they enter nature, they not only stay there, but also disperse to different parts of the environment, such as the atmosphere, water, soil, and living beings. Metal loads are enriched in soils, sediments, and living organisms through various unintentional and intentional processes [2]. HMs are derived from both natural and artificial sources. In the environment, they are induced by the weathering of metal, including rocks, volcanic eruptions, forest fires, and soil erosion [3]. These sources gradually but surely bring elements like As, Cd, Cr, copper (Cu), Pb, Hg, and zinc (Zn) to the soils and the natural waters. It is not so that natural sources are equally responsible for the pollution of HMs but a great deal of the HMs pollution has its roots in human activities such as mining and smelting operations, industrial effluents, fossil fuel combustion, waste disposal without proper treatment, vehicular emissions, wastewater from tanneries and electroplating industries, agricultural runoff containing fertilizers and pesticides, and sewage sludge application [4]. The increase of urbanization and the rapid industrialization have aggravated the issue of HMs release to the environment, increasing the chances of exceeding natural background levels and posing both ecological and health risks. They can be in the form of particulate matter (PM) and can be transported for a distance of several kilometers before they finally settle on the land or water bodies once released into the air [5]. Involving the metals in the waters, these metals get attached to the sediments or adsorbed by the water organisms, and therefore, the metals go through biomagnification in the food chains. They may also attach to the organic matter in the soils, thus changing their movement and bioavailability. The spread of HMs in the natural environment is supervised by a number of considerations that comprise pH, redox potential, organic matter content, and climate circumstances.

Heavy metals (HMs) build up over a very long period of time in natural habitats and can change the whole ecological system, cause major pollution of the water and soil sources, and seriously threaten the health of humans and animals alike [6]. Such metals have the ability to alter the diversity of living organisms and reduce the productivity of the ecosystem. Their existence negatively affects the entire environment. HMs are at the root of a range of environmental problems because they are highly persistent, can bioaccumulate in organisms, and, in addition, are toxic even at different trophic levels. The ultimate result of the accumulation and magnification of HMs along the food chain is the occurrence of various toxic effects in plants, animals, and humans [7]. In the case of humans, HM consumption through contaminated water, food, and air may cause neurodegenerative diseases, renal failure, cancer, and developmental problems, which are dependent on the nature and concentration of the metals [8]. They are able to penetrate biological barriers and have the potential to accumulate in essential organs, as well as cause genotoxic, neurotoxic, and carcinogenic effects [9]. For example, Pb infects the nervous system and slows down cognitive development in children, whereas Cd is the culprit of renal and bone diseases. As can be the origin of skin lesions and cancers, and Hg impairs neurological and reproductive functions [10].

Metal pollutants that are released into rivers and lakes can alter water chemistry and allow aquatic plants and animals to become endangered; thus, these aquatic bodies are the first to suffer from this kind of pollution [11]. Aquatic organisms such as algae, fish, and invertebrates are subjected to oxidative stress, enzymatic activity disruption, and reproductive dysfunction [12]. At the same time, even these harmful metals, at the lowest levels, are able to bring on long-term toxicity that may later affect the growth, behavior, and life continuation of both aquatic and land organisms. The lengthy biological half-life of many HMs exacerbates their health risks, thus making it necessary to have continuous monitoring and strict regulatory measures in place to protect the environment and human health. As a result, rigorous efforts are required to handle this problem, which should comprise measures for contamination stoppage, the usage of environmentally friendly knowledge for contamination elimination, and the implementation of rigorous protocols to control the issue of these contaminants and ensure the care of the atmosphere and the health of humans.

The requirement for less energy-consuming and more effective remediation approaches for HMs is due to the restrictions and environmental influences of traditional approaches, which are usually expensive, need a lot of energy, and cause the creation of secondary leftovers. Sustainable remediation is about the execution of environmentally friendly, low-priced, and resource-competent determines that have minimal or no negative effect on the environment while still being able to renew polluted ecosystems. Methods such as phytoremediation, bioremediation, and the use of biochar or nanomaterials (NMs) are in line with the principles of green chemistry and CE as they retain renewable reserves, lessen rubbish creation, and enable the long-term well-being of soil and water. These techniques not only guarantee the detoxification of HMs in an effective manner but also become a source of ecological balance and contribute to the realization of sustainable development goals (SDGs). In this study, we discuss the different approaches of HM remediation with the challenges and future perspectives, and also provide some recommendations.

6.2 Conventional Approaches to Heavy Metal (HM) Remediation

6.2.1 Physical Methods

Physically based remediation methods represent the most direct and historically known ways of eliminating or separating HMs from the pollution of natural surroundings [13]. Phase separation, sedimentation, filtration, and soil washing are examples of these procedures to separate or eliminate contaminants that have not been chemically modified. This also includes soil diggings, dredging, soil cleansing, thermal desorption, and electrokinetic remediation. As an illustration, soil cleansing utilizes water or chemical liquids to take out metal toxins, whereas dredging is the physical elimination of contaminated sediments from the environment. Such approaches are repeatedly selected for their proximity to impact and are very competent in the handling of extremely contaminated sites, in particular, those where contamination poses a risk to human health or the ecosystems. However, despite the fact that this method can effectively remove HMs, it still faces limitations related to problems of costs, site disturbance, and waste production. To illustrate the point, excavation and dredging can change the ecological balance of the scenery and produce a large volume of secondary waste that needs to be disposed of safely. Moreover, this method is usually more appropriate for contamination that is localized rather than that which is large-scale or spread out. It is typical for them to move pollutants from one medium to another instead of actually detoxifying or transforming them. As a result, the issue of whether this method is sustainable in the long run and environmentally friendly is raised quite frequently in contemporary remediation plans. To make physical remediation more effective and environmentally friendly, the use of this method is being combined with chemical or biological ones by researchers and practitioners.

One example is the combination of electrokinetic remediation with phytoremediation or chelation to enhance metal mobility and recovery rates [14]. Also, the stabilization steps may follow thermal desorption to prevent the release of contaminants again. These mixed solutions not only enhance the elimination efficacy to the extreme but also reduce the process budgets and the environmental impact. So, it is still fundamental that physical approaches be used for fast or alternative remediations, but they have to be combined with eco-friendly knowledge to attain environmental renovation and conform with existing environmental guidelines in the future.

6.2.2 Chemical Methods

Chemical remediation techniques like precipitation, oxidation-reduction, ion exchange, and neutralization are considered one of the most widely employed traditional techniques that are applied to get rid of HMs or to stabilize HMs in contaminated areas [15]. These practices include chemical reactions that change the metals into forms that are less soluble, less mobile, or less bioavailable. Chemical precipitation is a technique in which chemicals such as lime, sulfides, or carbonates are proposed to transform metal ions dissolved in water into mixtures that are insoluble and can be simply removed from the environment. Analogously, redox reactions can remove metals at distinct oxidation stages, thereby cleansing and reducing the movement of the metals in the environment. Moreover, there is a major metal removal chemical method, which is adsorption and ion exchange, where materials with particular surface characteristics or exchange capacities are employed to attract and retain metal ions. Activated carbon (C), zeolites, and artificial resins are commonly used for these purposes owing to their robust similarity for metal ions. Chelation is a popular chemical technique in which chelating mediators like DTPA (diethylenetriaminepentaacetic acid) and EDTA (ethylenediaminetetraacetic acid) are used in contaminated places to remove metals by forming a complex with metals [16]. While the use of artificial chelating mediators can generate secondary contamination, biodegradable alternatives are required. Although chemical remediation methods are effective and quite rapid, they are usually limited due to reagent costs, partial removal, and the possibility of environmental side effects. The issues of generating a huge volume of chemical sludge and the requirement for post-treatment disposal are the main disadvantages. Besides that, some chemical actions may alter the quality of the soil, thus changing its potency and microbial populations. However, chemical remediation is still noteworthy, and in combination with other techniques, such as biological or physical, it may be a more sustainable technique for HM remediation.

6.2.3 Limitations and Environmental Concerns of Conventional Methods

The conventional treatment technologies face substantial restrictions related to their effectiveness, price, and sustainability during the remediation of HMs. These technologies are high in the consumption of energy and need complex foundations. Besides, their efficiency at low metal concentrations is low, and most of the methods are non-selective, thus essential nutrients are removed together with toxic metals. What is more, the high operational and maintenance costs turn them into less viable options for large-scale or long-term applications in developing regions, especially in those areas where there is a lack of a reliable energy supply. Otherwise, the contaminants can be released back to the environment. In the same fashion, chemical reagents' utilization may bring new toxic compounds that change soil and water chemistry. Besides, some thermal or electrochemical methods operate on a substantial energy input that results in greenhouse gas (GHGs) emissions. The disadvantages of the traditional methods emphasize the importance of finding new, environmentally friendly, inexpensive, and sustainable methods that will restore not only the polluted sites but also the planet.

6.3 Sustainable and Green Approaches to Heavy Metal (HM) Remediation

Green remediation principles aim to reduce the environmental impact of the remediation activities, as the contaminant removal is efficient. This strategy involves the use of renewable resources, the reduction of waste, the saving of energy, and the protection of the ecosystem at each stage of the remediation process. Among the principles, there are measures to decrease the release of GHGs, save water and energy, avoid the occurrence of pollution as a result of the process, and support the reuse and recycling of materials. Also, green remediation supports the use of the LCA in order to consider the environmental effects of different options and to be sure that the chosen technology not only cleans HMs but also keeps nature and is sustainable in the long run. When discussing environmentally friendly and green solutions for HM remediation, the strategies are supposed to consider the three pillars of sustainability, i.e., the environmental, economic, and social aspects. The employment of phytoremediation, bioremediation, and biochar or NMs made from renewable resources is one of the instances that demonstrate this model. These are natural biological and physicochemical processes that the methods utilize to bind, remove, or change (convert) toxic metals in soils and water without the release of harmful by-products. By concentrating on low-impact, cheap, and resource-saving technologies, green remediation as a part of sustainable development serves as a way to pollute less, thus being a double-winner solution for both the environment and the local community. Such approaches to local HM contamination are eco-friendly, economically viable, socially acceptable, and ethically sound.

Sustainability metrics together with LCA are, first of all, the means by which we understand the extent and, secondly, the ways of improving sustainable and green methods for HMs removal. They allow forming an integrated framework for quantifying the environmental, economic, and social impacts associated with different types of remediation technologies throughout their entire life cycle, i.e., starting from raw materials acquisition, through energy consumption and installation, to waste generation and end-of-life management. LCA achieves this by looking at a comprehensive set of parameters like C footprint, energy use, resource consumption, and potential for toxicity, thus allowing the identification of trade-offs and hotspots in remediation processes. The employment of sustainability metrics ensures that remediation methods become not only those that effectively remove pollutants but also ones that cause minimal ecological disturbance, enable resource recovery, and are congruent with the CE principles, thus they help in making decisions about the most environmentally friendly and cheapest solutions for HM contaminated sites. Highlighting the concepts of resource recovery, waste reduction, and environmentally friendly reuse of resources within the frameworks of remediation, integrating CE concepts with the HM clean-up is quite effective. The CE-supported remediation methods are aimed at altering the polluted materials into significant resources with the help of advanced knowledge. As a solution, the recycled metals can be reused in the manufacturing process. So, this process can reduce the time and need for new materials. Furthermore, the CE model can help to achieve the sustainability targets with lower emissions of GHGs and reduce the generation of contamination. Figure 6.1 briefly summarizes the various types of HMs remediation strategies that could be employed.

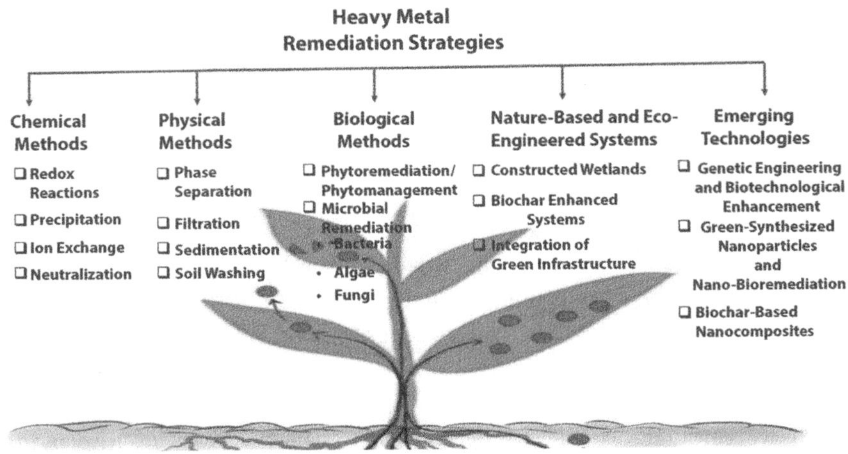

Figure 6.1 Different types of heavy metal (HM) remediation strategies.

6.3.1 Biological Methods

6.3.1.1 Phytoremediation/Phytomanagement

Phytoremediation is an economically effective and environmentally friendly method that uses plants to clean toxins from the surroundings [17]. Certain plant groups have the ability to absorb elevated concentrations of HMs in their tissues, and these plants are identified as hyperaccumulators [18]. This technique is not only able to detoxify the HM contaminated soils but also to consolidate and freshen the lands that are desolate, while at the same time producing the co-benefits that are ecological and economic. Phytoextraction, phytostabilization, rhizofiltration, and phytovolatilization are the mechanisms by which plants can remove HMs from polluted sites and immobilize HMs to reduce their bioavailability. This technique is a sustainable alternative approach to the conventional technique. It restores the ecological balance of the polluted sites and minimizes the generation of secondary pollution. Furthermore, it offers multiple benefits with the concept of CE principles. The biomass generated during the remediation of HMs can be used for bioenergy production and the recovery of valuable metals. The integration of this technique with microbial-supported processes extends and improves metal uptake and detoxification efficacy through upgraded nutrient cycling and rhizospheric action. While this technique is restricted due to the growth rate of plants, root length, and pollutants' bioavailability, continuing improvements in advanced techniques, such as genetic engineering and biotechnology, improve the accumulation capacity of plants and improve this approach as a sustainable and low-impact approach for HM remediation. Table 6.1 provides a list of different plants for the removal of HMs.

6.3.1.2 Microbial Remediation

Microbial remediation is an eco-friendly and green technique. Bacteria, algae, and fungi remediate HMs from contaminated sites through sorption, accumulation, transformation, and mineralization mechanisms. *Pseudomonas, Azotobacter, and Bacillus* are some species of bacteria that have metal-binding proteins and remove HMs in a sustainable approach [62]. *Aspergillus, Trametes, and Penicillium* are species of fungi that have a great surface area and unique functional groups. These fungi are effective in absorbing HM ions on their cell walls, and this procedure is known as mycoremediation [63]. Glucans, proteins, and chitin are present in the cell wall of the fungi that increase the efficiency of fungi to bind metal ions and remove HMs from polluted soils, sediments, and wastewater. The fungi release organic acids and enzymes and convert the metals into less toxic or more stable forms. This remediation method has far-reaching benefits compared to the conventional methods since it works under mild conditions and lessens the secondary pollution generation. The method can also be used *in situ* to revive the polluted ecosystems and keep soil fertility and ecological balance. Besides, genetically engineered microorganisms (GEMs) and microbial consortia have been developed to enhance metal removal efficiency in complex pollution situations. The adoption of this remediation method with other sustainable methods like phytoremediation and biochar can elevate the capacity for long-term HM remediation. Essentially, this remediation technique is a stepping stone towards the implementation of sustainable remediation methods, which are in line with the principles of green chemistry and circular bioeconomy. Table 6.2 summarizes different microbes and their bioremediation potential for HMs.

6.3.2 Nature-Based and Eco-Engineered Systems

6.3.2.1 Constructed Wetlands and Assisted Natural Attenuation (ANA)

Constructed wetlands stand out as a highly efficient nature-based and eco-engineered solution for the removal of HMs from contaminated waters and wastewater systems [109]. These engineered systems are designed to

Table 6.1 List of different plants for the removal of heavy metals (HMs).

S. no.	Name of the plants	Heavy metal (HM)	References
1	*Eleocharis acicularis*	As	[19]
2	*Tagetes minuta*	As and Pb	[20]
3	*Lepidium sativum* L.	As, Cd, and Pb	[21]
4	*Azolla pinnata*	Cd	[22, 23]
5	*Cannabis sativa* L.		[24]
6	*Thlaspi caerulescens*		[25]
7	*Sorghum bicolor* L.	Cd, Cu and Zn	[26]
8	*Rapanus sativus* L.	Cd, Fe, Pb, and Cu	[27]
9	*Cicer aeritinum* L.	Cd, Pb, Cr, and Cu	[28]
10	*Eichhornia crassipes* L.	Cr and Zn	[29]
11	*Haumaniastrum katangense*	Cu	[30]
12	*Oryza sativa* L.	Cu and Cd	[31]
13	*Miscanthus x giganteus*	Cu, Ni, Pb, and Zn	[32]
14	*Lactuca sativa* L.	Cu, Mn, Zn, Ni and Cd	[28]
15	*Brassica juncea* L.	Cu, Zn, and Pb	[33]
16	*Brassica napus* L.		[34]
17	*Spinacia oleracea* L.	Cu, Ni, Zn, Pb, and Cr	[35]
18	*Jaltropa curcas* L.	Cu, Mn, Cr, As, Zn, and Hg	[36]
19	*Achillea millefolium*	Hg	[37]
20	*Marrubium vulgare*		[38]
21	*Pteris vittata*		[39]
22	*Alyssum murale*	Ni	[40]

(*Continued*)

Table 6.1 List of different plants for the removal of heavy metals (HMs). (*Continued*)

S. no.	Name of the plants	Heavy metal (HM)	References
23	*Brassica nigra*	Pb	[41]
24	*Betula occidentalis*		[42]
25	*Cucumis sativus* L.		[43]
26	*Lantana camara* L.		[44]
27	*Lavadula vera* L.		[45]
28	*Lens culunaris Medic.*		[46]
29	*Medicago sativa*		[47]
30	*Noccaea Caerulescens*		[48]
31	*Viola principis*		[49]
32	*Sorghum halepense* L.		[50]
33	*Minuartia verna, Agrostis tenius*		[51]
34	*Pelargonium Sp.*		[52]
35	*Helianthus annuus*	Pb and Cd	[53]
36	*Salvia sclarea* L.	Pb, Cd, and Zn	[54]
37	*Euphorbia cheiradenia*	Pb, Zn, Cu, and Ni	[30]
38	*Pisum sativum* L.	Pb, Cu, Zn, Ni, As, and Cr	[55]
39	*Potentila griffithii*	Zn	[56]
40	*Arabidopsis halleri*		[57]
41	*Thalaspi caerulescens*		[58]
42	*Brassica oleracea, Raphanus sativus*	Zn, Cd, Ni and Cu	[59]
43	*Trifolium alexandrinum*	Zn, Pb, Cu, and Cd	[60]
44	*Cardaminopsis halleri*		[61]

Table 6.2 List of different microbes and their bioremediation potential for heavy metals (HMs).

S. no.	Microbial group	Species name	Heavy metals (HMs)	References
1	Bacteria	*Bacillus* sp., *Microbacterium sp., Micrococcus sp.,* and *Shigella sp.*	As and U	[64]
		Staphylococcus hominis AMB-2	Cd and Pb	[65]
		Sphingomonas paucimobilis 20006FA	Cr	[66]
		Pannonibacter phragmitetus BB		[67]
		Bacillus sp. MNU16		[68]
		Micrococcus luteus DE2008	Cu and Pb	[69]
		Bacillus strain MRS-2 (ATCC 55674)	Pb	[70]
		Staphylococcus aureus		[71]
		Pseudomonas aeruginosa	Hg	[72]
		Sphingobium sp. SA2		[73]
		Bacillus megaterium and *Rhizopus stolonifera*	Ni and Cd	[74]
2	Fungi	*Aspergillus flavus*	As	[75]
		Penicillium sp.	Cd, Pb, and Hg	[76]
		Trichoderma sp.	Cd	[77]
		Ganoderma lucidum	Cd and Pb	[78]
		Aspergillus niger	Cr	[79]

(Continued)

Table 6.2 List of different microbes and their bioremediation potential for heavy metals (HMs). (*Continued*)

S. no.	Microbial group	Species name	Heavy metals (HMs)	References
		Penicillium sp., *Trichoderma* sp., and *Aspergillus* sp.	Co and Cu	[80]
		Fusarium sp.	Pb, Cr, and Cu	[81]
		Aspergillus nomius JAMK1	Pb, Cu, and Ni	[82]
		Saccharomyces cerevisiae	Ni, Hg, and Pb	[83]
3	Algae	*Ulva Reticulata*	As	[84]
		Spirogyra hyaline	Cd, Hg, Pb, and As	[85, 86]
		Phormidium ambiguum	Cd	[87]
		Cladophora glomerata and *Enteromorpha intestinalis*	Cr	[88]
		Enteromorpha intestinalis		[89]
		Chlorococcum humicola	Co	[90]
		Chlorophyceae sp.	Cu	[91]
		Desmodesmus sp. AARLG074		[92]
		Oscillatoria quadripunctulata,	Cu and Pb	[59]
		Cladophora fascicularis	Pb	[93, 94]
		Sargassum natans		[95, 96]
		Chlorella vulgaris	Mn	[91]

(*Continued*)

Table 6.2 List of different microbes and their bioremediation potential for heavy metals (HMs). (*Continued*)

S. no.	Microbial group	Species name	Heavy metals (HMs)	References
		Phormidium ambiguum	Hg	[87]
		Ulva lactuca		[97]
		Oedogonium westi	Ni	[98]
		Ascophyllum nodosum	Ni and Pb	[95, 99]
		Chlorophyceae sp.	Zn	[91]
		Fucus vesiculosis, Laminaria japonica		[100]
4	Yeast	*Candida tropicalis*	Cr	[101]
		Saccharomyces cerevisiae		[102]
		Gelidium amansii	Pb	[103]
5	Microbial consortia	*Antrodia serialis, Paecilomyces lilacinus,* and *Penicillium cataractum*	As, Fe, Cu, Mn, and Cr	[104]
		Daldinia starbaeckii, Perenniporia subtephropora, Phanerochaete concrescens, Fusarium equiseti, Cerrena aurantiopora, Polyporales sp., Aspergillus fumigatus, Trametes versicolor, and *Aspergillus niger*		[105]
		Saccharomyces cerevisiae and Pseudomonas aeruginosa	Cr	[106]

(*Continued*)

Table 6.2 List of different microbes and their bioremediation potential for heavy metals (HMs). (*Continued*)

S. no.	Microbial group	Species name	Heavy metals (HMs)	References
		Pseudomonas aeruginosa and Bacillus subtilis		[102]
		Streptomyces sp. M7, *Streptomyces* sp. MC1, *Streptomyces* sp. A5, and *Amycolatopsis tucumanensis*		[105]
		Acinetobacter sp. and *Arthrobacter* sp.		[107]
		Saccharomyces cerevisiae and Bacillus subtilis		[102]
		Penicillium sp. A1 and *Fusarium* sp. A19	Cr, Pb, Cu, and Zn	[107]
		Pseudomonas pyogenes, Serratia marcescens, Erwinia amylovora, and *Enterobacter cloacae*	Pb, Cr, As, Ni, and Zn	[108]
		Bacillus subtilis and *Staphylococcus aureus*		

resemble natural wetlands both in their structure and function by combining vegetation, substrates, and microbial communities that are capable of removing pollutants through their physical, chemical, and biological processes working together. HMs are basically fixed in such an environment through sedimentation, filtration, adsorption onto substrate materials, plant uptake, and microbial conversions like sulfate reduction and precipitation. The use of certain plants, e.g., *Typha latifolia, Phragmites australis,* and *Scirpus* spp., not only helps the metal uptake and stabilization but also ensures the metal safety, while gravel, sand, and organic matter serve as adsorbing agents and ion exchangers.

These wetlands are, from a sustainability point of view, an inexpensive, low-energy, and environmentally friendly alternative to conventional HM treatment technologies. Besides that, they not only supply the ecosystem with the services that it lacked before, like biodiversity enhancement, C sequestration, and the improvement of the landscape by its aesthetic, but also are a perfect solution. The adaptability of these wetlands to diverse environmental requirements and their possible usage of CE principles, such as the reprocessing of biomass for bioenergy or biochar manufacture, thus substantially increases their environmental and economic profits. Their widened, long-term performance will, however, depend on the applicable design, hydraulic administration, and repeated conservation to check metal collection saturation. These wetlands correspond to a wonderful example of a green and multifunctional system solution to HM contamination that also facilitates ecological resilience and resource recovery.

Assisted Natural Attenuation (ANA) is a method that, directly and massively, helps phytomanagement through the facilitation of various natural biogeochemical processes that decrease the toxicity, mobility, and bioavailability of HMs in soils and groundwater [110]. The method operates on the principle of minimal human intervention and relies primarily on naturally occurring physical, chemical, and biological processes such as adsorption, precipitation, complexation, and microbial transformation. However, the eco-engineering modifications, such as nutrient additions, organic substrate, or a specific plant species, may be employed to increase these natural attenuation pathways in the ANA. This low-energy and inexpensive intervention is an area of particular suitability for large or low-risk contaminated sites, thus enabling them to recover slowly while still maintaining ecological balance and sustainability.

6.3.2.2 *Biochar-Enhanced Systems*

Biochar is an excellent material to lift a system and is a nature-inspired, eco, human-made solution that has been developed to localize HM pollution in soils and waters [111]. It is a C-rich material gained by pyrolyzing biomass under restricted oxygen (O_2) requirements. It has a particularly great external zone, pore volume, and an excessive number of functional groups that can adsorb, complex, and restrain HMs. The addition of biochar to metal-contaminated sites not only stabilizes the metals but also makes them less mobile and less bioavailable. Besides that, biochar can change soil pH, cation exchange capacity, and microbial activity, which are factors that determine the performance of the remediation process. Its binding to different metals such as Pb, Cd, Cr, Cu, and Zn has been extensively studied, so biochar can be considered as a green and cheap alternative to the traditional sorbents.

Systems enriched with biochar do not end at simply passive adsorption and can be intentionally mixed with other remediation methods like phytoremediation, microbial remediation, and constructed wetlands to yield synergistic effects. To illustrate the point, biochar-treated soils become not only a medium for plant growth and root development but also an agent that enhances phytoextraction and metal stabilization. Similarly, biochar in wetland systems is a reactive substrate that elevates pollutant retention and, at the same time, is a habitat for the microorganisms that are beneficial and involved in biotransformation processes. The lifespan, C sequestration potential, and renewable nature of biochar are consistent with the principles of the CE and sustainable remediation. Hence, biochar-based, engineered systems are a fully integrated, low-impact, and scalable approach to HM contamination of the earth and water bodies over the long term.

6.3.2.3 *Integration of Green Infrastructure for Urban Heavy Metal (HM) Mitigation*

The deployment of green infrastructure in cities provides a viable and eco-friendly approach to lessen HM contamination. Green infrastructure components, e.g., green roofs, urban forests, bioswales, rain gardens, and permeable pavements, are essentially the filters that take in and keep HMs from urban stormwater runoff and atmospheric deposition [112]. The plant and soil parts of such systems can chemically bind and stabilize metals like Pb, Cd, Cu, and Zn; thus, they do not allow them to be discharged to aquatic systems and groundwater. Furthermore, the application of metal-tolerant plants improves phytostabilization and phytoextraction platforms; hence, their solutions for forever decontamination. Moreover, these multifunctional drivers of contamination also upgrade urban air excellence, microclimate adaptation, and biodiversity; thus, they link environmental sanitization with the basis of urban sustainability. Eco-engineered features in green infrastructure systems provide them with a greater capacity for HM reduction by adjusting hydrological flow paths, substrate composition, and plant-microbe interactions. As an illustration, biochar, compost, and mineral additives are capable of raising the metal sorption capacity of the soil and reducing the metal bioavailability, while microbial consortia that are injected into the soil facilitate biotransformation and immobilization of metals. The use of sensor-based monitoring and smart water management integration also guarantees system efficiency and resilience to varying pollution loads. Green infrastructure, which combines natural processes with engineering design, is a money-saving, low-C alternative to the use of conventional stormwater management and remediation technologies; thus, it is contributing to the return to nature-based urban solutions for HM pollution control.

6.3.3 Emerging Technologies for Heavy Metal (HM) Remediation

6.3.3.1 Genetic Engineering and Biotechnological Enhancements

The efficiency of bioremediation techniques for HM remediation has substantially improved with the advancement in genetics and biotechnology. Metal-binding proteins are present in transgenic plants, which makes these plants capable of removing HMs from the soil and water [113]. Similarly, metal-binding specific enzymes are present in recombinant bacteria, and these enzymes can enhance the capability to remove HMs [114]. The microbial strains with improved metabolic pathways for biosorption, bioaccumulation, and biotransformation of HMs have been developed through CRISPR-Cas9 and gene cloning techniques. By these microbial strains, the specificity and speed of the remediation process have been increased. Metagenomics, transcriptomics, and proteomics are the biotechnological tools that have increased our understanding of microbial populations involved in HM remediation. These tools help researchers to find key enzymes, metabolic pathways, and genes for the remediation of HMs. The incorporation of synthetic biology assists in the design of microbial consortia by balancing functions and improving the performance and stability under different environmental settings. The use of bioinformatics models can help in the prediction of microbial behavior and optimize the efficiency of the remediation of HMs. At the end, we can say that these techniques act as an eco-friendly and sustainable approach to managing HM pollution and reducing the dependency on energy and chemical-concentrated remediation techniques. Table 6.3 highlights some GEMs with the bioremediated HMs.

6.3.3.2 Green-Synthesized Nanoparticles (NPs) and Nano-Bioremediation

Green-synthesized NPs are made through plant extracts or microbial routes [133]. These particles not only controlled morphology but also enhanced surface reactivity. They have provided a sustainable and environmentally

Table 6.3 List of various genetically engineered microbes (GEMs) with the bioremediated heavy metals (HMs).

S. no.	Microbial species	Heavy metals (HMs)	Genetic modification	Enhancement of outcomes	References
1	*Saccharomyces cerevisiae*	As	WaarsM gene expression	Through volatization, tolerance for As is improved	[115]
2	*Bacillus Idriensis* and *Sphingomonas desiccabilis*		arsM gene overexpression	Enhances As removal through the process of biovolatilization in soil as well as water environments	[116]
3	*Escherichia coli*		Simultaneously, expression of fMT and As transporter GlpF	Increase As accumulation	[117]
4	*Bacillus subtilis* BR151 (pTOO24)	Cd	Luminescent Cd sensors	Increased bioavailability of Cd in soil	[118]
5	*Caulobacter crescentus JS4022/ p723-6H*		RsaA-6His fusion protein	Enhances the retrieving capacity of Cd in water	[119]
6	*Pseudomonas putida* 06909		Metal-binding peptide EC20 expression	Improved the binding of Cd, as well as alleviated the cellular toxicity of Cd	[120]
7	*Mesorhizobium huakuii* B3		Expression of the gene Phytochelatin synthase (PCS)	Increase the accumulation of Cd	[121]

(*Continued*)

Table 6.3 List of various genetically engineered microbes (GEMs) with the bioremediated heavy metals (HMs). (*Continued*)

S. no.	Microbial species	Heavy metals (HMs)	Genetic modification	Enhancement of outcomes	References
8	*Pseudomonas putida*	Cr	ChrR overexpression	Increased reduction of Cr	[122]
9	*Methylococcus capsulatus*		CrR genes for Cr (VI) reductase activity	Increased Cr bioremediation	[123]
10	*Pseudomonas fluorescens* OS8; *Escherichia coli MC1061; Bacillus subtilis BR151 and Staphylococcus aureus RN4220*	Cd, Hg, and Zn	MerR/CadC/ZntR/Pmer/PcadA/PzntA	Enhanced Cd, Zn, and Hg bioavailability in soil	[124]
11	*Acidithiobacillus ferrooxidans*	Hg	Gene expression of Hg ion transporter	Increased potential in the uptake of Hg	[125]
12	*Escherichia coli*		Transposon Tn21 encoding MerE protein	Increases Hg uptake	[126]
13	*Achromobacter* sp. AO22		Hg reductase expressing mer gene	Improved Hg *in-situ* bioremediation	[127]
14	*Escherichia coli* JM109		Metalloregulatory protein MerR using an INP anchor	Enhanced Hg biosorption	[128]
15	*Deinococcus radiodurans*		Hg (II) resistance gene (merA)	Significantly reduces Hg at higher temperatures as well as at ionizing radiation	[129]
16	*Escherichia coli*		Expression of polyphosphate kinase and metallothionein	Increases accumulation as well as tolerance of Hg	[130]

(*Continued*)

Table 6.3 List of various genetically engineered microbes (GEMs) with the bioremediated heavy metals (HMs). (*Continued*)

S. no.	Microbial species	Heavy metals (HMs)	Genetic modification	Enhancement of outcomes	References
17	*Escherichia coli* SE5000	Ni	Ni transport system and metallothionein	Improved uptake of Cd from aqueous solutions	[131]
18	*Escherichia coli*	Ni and Co	Serin acetyltransferase overexpression	Improves resistance against Ni as well as Co	[132]

friendly result for the elimination of HMs; thus, they characterize the incorporation of nanotechnology (NT) with the values of green chemistry. They decrease the incidence of secondary pollution and promote biocompatibility. In comparison with the volume, their small size and modifiable physicochemical properties render them very efficient in the removal of trace metals from soils, sediments, and wastewater. For instance, the plant extract used in the synthesis of iron (Fe), zinc oxide (ZnO), and silver (Ag) NPs can make these NPs have an extremely high metal-adsorbing capacity, by the mechanisms of surface complexation and redox reactions, thus providing a safe environment for the removal of HMs. In contrast to the conventional physical or chemical methods of synthesis that usually require the use of toxic reagents and high energy, green synthesis makes use of biological entities such as plants, bacteria, fungi, or algae as reducing and stabilizing agents. These NPs can be employed for the efficient adsorption, reduction, or transformation of metal ions in contaminated matrices. The application of these NPs in composite materials, biofilters, or membranes leads to their stabilization and reuse; hence, they are envisaged for large-scale operations. The recent surveys suggest the recycling of agro-leftover or microbial biomass for NP creation, thus opening the way from trash valorization to trash control.

Nano-bioremediation is a technology that merges the advantages of NTs with those of biological remediation to clean and detoxify HMs from the environment in a very effective way [134]. In this combined method, NPs like zero-valent Fe, titanium dioxide (TiO_2), or magnetite (Fe_3O_4) are combined with microorganisms, enzymes, or plants for metal bioavailability, adsorption, and transformation processes. The nanoscale materials significantly increase the surface area, catalytic activity, and redox potential, thus interaction between pollutants and biological agents becomes faster and more efficient. For instance, zero valent Fe particles can perform a reduction of toxic metal ions (e.g., Cr (VI) to Cr (III)) while microbial biofilms can immobilize or even metabolize the less toxic species that have been thus generated, detoxification thereby being renewable and eco-friendly. Such hybrid technology is less harmful to the environment and, therefore, can serve as a green option to the usual physicochemical treatments. Besides not polluting the environment again, it also lessens the chemical use and makes it possible to apply the treatment locally in mild environmental conditions. Green, synthesized NPs also help nano-bioremediation to be in line with the sustainability principles by lessening the toxicity and increasing biodegradability. The technologies combined in the hybrid approach herald a new era of clean water and a sustainable environment. The combined use of NMs and biological agents in metal-contaminated soils, sediments, and wastewater is among the most studied and confirmed applications in this area. However, to entirely recognize the contact between these small structures and living organisms, we want to identify the precise quantities and still keep the environment harmless; thus, extra effort has to be made before we can use this knowledge extensively and in a sustainable manner for environmental administration.

6.3.3.3 Biochar-Based Nanocomposites

Biochar-based nanocomposites can be considered as a brilliant concept of NT and environmentally friendly materials that are suitable to be used for the removal of HMs pollution [135]. The nanocomposites that are made by combining biochar with NPs such as metal oxides (Fe_3O_4, TiO_2, and ZnO), graphene, or zero valent metals have improved sorption capacity, catalytic activity, and redox potential. In these materials, the adsorption capability of biochar and the high reactivity of NPs are combined synergistically; thus, the toxic metal species such as Pb (II), Cd (II), Cr (VI), and As (III) that are released into the environment from soils and waters are either tightly bound or converted to less toxic forms. Additionally, the magnetic properties conferred by some NPs make them easily recovered and reused, thus their economic and operational viability in environmental clean-up is increased.

The fabrication of biochar-based nanocomposites *via* green routes is in perfect harmony with the sustainable remediation objectives, as it entails fewer chemical inputs, the use of waste biomass, and the absence of secondary pollution. Functionalization methods, for instance, impregnation, co-precipitation, and sol-gel methods, can be carried out under gentle conditions with the use of eco-friendly reagents, thus preserving the environment. Besides, these substances have excellent stability and can be reused; thus, through the principles of the CE, they can be used for several cycles of remediation. Moreover, the use of biochar and nanocomposites is not limited to adsorption, but they can also be involved in photocatalytic degradation and redox-mediated detoxification of metals, thus, their application potential is extended. In summary, biochar-based nanocomposites represent a sustainable, economical, and scalable solution to HM pollution, which is a fusion of NT innovations with green material science for the protection of the environment in the long run.

6.4 Socioeconomic and Policy Dimensions of Sustainable Remediation

The sustainable remediation socioeconomic aspects emphasize the close connection of nature restoration with the economic and social progress of the society. In order for HM remediation to be successful, the strategy should involve the technical side, community participation, livelihood, and economic sustainability of the region. Sustainable remediation, through using local knowledge and creating jobs *via* green technologies, can become a vehicle for inclusive development in areas where contamination from the industry has caused the communities to be left behind. The transition to eco-innovative and affordable technologies of remediation, like phytoremediation and biochar-based systems, is a step towards economic relief while at the same time making sure that the efforts lead to social equity and resilience. Considering the policy angle, sustainable remediation is dependent on well-planned frameworks that fuse the environmental, economic, and social goals into the existing governance structures.

The policymakers should deploy tiered strategies, from local regulations to international agreements, which not only stimulate cleaner production, implement the polluter pays principle (PPP), but also encourage CE practices. Proper policy tools, such as grants for green technologies, tax breaks for pollution control, and tight regulations on the release of industrial effluents, can motivate pollution abatement and the innovation spirit. In addition, the environmental standard and monitoring protocol synchronization are vital for ensuring accountability is secured, and technology transfer among the countries is made easier. Engaging the public and incorporating the policy aspect are two of the most essential factors for achieving long-term success in the cleanup of HMs. Through participatory decision-making processes, communities living in metal-contaminated areas are given the power to influence the priorities of the clean-up and thus guarantee the fair distribution of the results. Improving the capacity of institutions and creating the possibility of collaboration between the different sectors, government agencies, research institutions, and private enterprises can lead to the fast implementation of eco-friendly solutions. Moreover, the integration of sustainability measures, such as LCA, cost-benefit analysis, and social impact evaluation, into policy frameworks enables the protection of the environment to be in harmony with its economic aspect. These socioeconomic and policy proportions, when merged, establish the core of natural remediation; thus, they are the means through which environmental reclamation is in line with the larger societal intentions of integrity, strength, and success.

6.5 Challenges, Gaps, and Future Perspectives

Though substantial improvement has been achieved, several challenges and gaps in knowledge remain, which still slow down the large-scale treatment of environmentally friendly techniques for HM remediation. The main challenge is the variety of environmental circumstances that affect the effectiveness of biological and nature-based practices, using soil pH, natural matter contents, and metal speciation. Moreover, the insufficient knowledge of the stability and fate of immobilized or transformed metals in the long term raises questions about the potential re-mobilization of metals under different environmental conditions. In addition to that, economic and infrastructural constraints, combined with the lack of standardized protocols for assessment and regulatory frameworks, make it difficult to go from laboratory innovations to the field scale. There is also a concern regarding the use of NT in bioremediation; the potential ecotoxicity of NMs is one of the reasons why their deployment should be thoroughly assessed before a wide application.

The following perspectives emphasize the need for thorough and interdisciplinary strategies that entail biotechnological, NM, and nature-based innovations within the framework of a CE. One of the definite ways to conclude the efficiency and sustainability levels is the development of mixed remediation approaches that combine physical, chemical, and biological methods. The improvements in omics knowledge, artificial intelligence (AI), and remote sensing can be tied together to get an in-depth knowledge of pollutant dynamic range and to make remediation policies more effective in real time. Likewise, the application of global strategies, green certification proposals, and community-driven strategies will be needed to accelerate the acceptance of environmentally friendly remediation methods. Eventually, the forthcoming attempts should be focused not only on the elimination of HMs but also on the renewal of ecosystem acts and the upgrade of source recovery that would lead to a truly sustainable environmental organization model.

6.6 Recommendations

The future attempts should highlight integrated, low-impact methods that merge natural, nanotechnological, and nature-based organizations to enhance sustainable approaches for HM remediation. Biochar composites, green-synthesized NPs, and plant microbe consortia are cost-effective, scalable, and eco-friendly materials, and these materials should be emphasized to develop. Scientists and policymakers should focus on partnership and encourage the values of the CE. Vigorous supervising frameworks should be applied to estimate the long-term ecological protection and effectiveness. Additionally, the participation of stakeholders, community awareness, and supportive strategy procedures are required globally to ensure the adoption and implementation of green remediation practices.

6.7 Conclusion

Sustainable methods for HM remediation represent a shift from energy-intensive, conventional, and environmentally harmful methods towards resource-efficient, eco-friendly, and socially responsible alternatives. These sustainable methods detoxify polluted environments and promote ecosystem resilience and circular resources. Phytoremediation, biochar application, microbial consortia, and green-synthesized NMs are examples of some methods that show how natural and engineered systems can work together to achieve remediation efficacy as well as sustainability. Furthermore, assessments of the life cycle and green metrics help make sure that these interferences decrease the generation of secondary pollution and carbon footprints and align the remediation technologies with the goal of sustainability worldwide. Generally, the HM remediation ability is positioned in interdisciplinary discovery that links environmental science, biotechnology, material engineering, and socio-economic strategy. Constant and long-term investigation is needed to examine the long-term effects of sustainable methods for their practical application. The collaboration between the university, manufacturing, and policymakers can execute the research laboratory-level investigation at the field level and confirm environmental restoration that is both working and equitable. Eventually, we can determine that sustainable remediation approaches suggest a pathway toward safer civilizations, a cleaner environment, and a circular and resilient environmental organization system.

List of Abbreviations

Ag	Silver
AI	Artificial intelligence
ANA	Assisted Natural Attenuation
As	Arsenic
C	Carbon
Cd	Cadmium
CE	Circular economy
Co	Cobalt
Cr	Chromium
Cu	Copper
DTPA	Diethylenetriaminepentaacetic acid
EDTA	Ethylenediaminetetraacetic acid
Fe	Iron
Fe_3O_4	Magnetite
GEMs	Genetically engineered microorganisms
GHGs	Greenhouse gas
Hg	Mercury
HMs	Heavy metals
LCA	Life cycle assessment
Mn	Manganese
Ni	Nickel
NMs	Nanomaterials
NPs	Nanoparticles
NT	Nanotechnology
O_2	Oxygen
Pb	Lead
PCS	Phytochelatin synthase
PM	Particulate matter
PPP	Polluter Pays Principle
SDGs	Sustainable development goals
TiO_2	Titanium dioxide
U	Uranium

Zn	Zinc
ZnO	Zinc oxide

AI disclosure statement: The author acknowledges the use of AI-based tools for minor language editing or grammatical correction. However, all intellectual contributions, analyses, and conclusions presented in this chapter are entirely the author's own.

References

1. Bhardwaj, L.K., Sharma, S., Jindal, T., Estimation of physico-chemical and heavy metals in the lakes of Grovnes & Broknes Peninsula, Larsemann Hill, East Antarctica. *Chem. Afr.*, 6, 5, 2677–2694, 2023.
2. Jadaa, W. and Mohammed, H., Heavy metals–definition, natural and anthropogenic sources of releasing into ecosystems, toxicity, and removal methods–an overview study. *J. Ecol. Eng.*, 24, 6, 249–271, 2023.
3. Garrett, R.G., Natural sources of metals to the environment. *Hum. Ecol. Risk Assess.*, 6, 6, 945–963, 2000.
4. Alam, S., Bhardwaj, L.K., Mallick, R., Rai, S., Estimation of heavy metals and fluoride ion in vegetables grown nearby the stretch of River Yamuna, Delhi (NCR), India. *Indian J. Environ. Prot.*, 43, 1, 64–73, 2023.
5. Popoola, L.T., Adebanjo, S.A., Adeoye, B.K., Assessment of atmospheric particulate matter and heavy metals: a critical review. *Int. J. Environ. Sci. Technol.*, 15, 5, 935–948, 2018.
6. Alengebawy, A., Abdelkhalek, S.T., Qureshi, S.R., Wang, M.Q., Heavy metals and pesticides toxicity in agricultural soil and plants: Ecological risks and human health implications. *Toxics*, 9, 3, 42, 2021.
7. Ali, H. and Khan, E., Trophic transfer, bioaccumulation, and biomagnification of non-essential hazardous heavy metals and metalloids in food chains/webs—Concepts and implications for wildlife and human health. *Hum. Ecol. Risk Assess.: Int. J.*, 25, 6, 1353–1376, 2019.
8. Chowdhury, F.N. and Rahman, M.M., Source and distribution of heavy metal and their effects on human health, in: *Heavy metal toxicity: human health impact and mitigation strategies*, pp. 45–98, Springer Nature Switzerland, Cham, 2024.
9. Bhardwaj, L.K., Rath, P., Choudhury, M., Metzdorff, A., Jodder, P.K., Electronic wastes and contamination of water, in: *Electronic Waste*, pp. 48–57, CRC Press, United Kingdom, 2025.
10. Bhardwaj, L.K., Rath, P., Upadhyay, D., Singh, V., Arsenic Removal from the Environment Using Biological Methods, in: *Emerging and Innovative Arsenic Removal Technologies for a Sustainable Future*, pp. 129–142, Springer Nature Switzerland, Cham, 2025.

11. Sharma, M., Kant, R., Sharma, A.K., Sharma, A.K., Exploring the impact of heavy metals toxicity in the aquatic ecosystem. *Int. J. Energy Water Resour.*, 9, 1, 267–280, 2025.
12. Lushchak, V.I., Environmentally induced oxidative stress in aquatic animals. *Aquat. Toxicol.*, 101, 1, 13–30, 2011.
13. Peng, W., Li, X., Xiao, S., Fan, W., Review of remediation technologies for sediments contaminated by heavy metals. *J. Soils Sediments*, 18, 4, 1701–1719, 2018.
14. Cameselle, C., Chirakkara, R.A., Reddy, K.R., Electrokinetic-enhanced phytoremediation of soils: status and opportunities. *Chemosphere*, 93, 4, 626–636, 2013.
15. Patterson, J.W. and Minear, R., Physical-chemical methods of heavy metals removal, in: *Heavy metals in the aquatic environment*, pp. 261–276, Pergamon Press (an imprint of Elsevier), Elsevier Ltd. Oxford, UK, 2013.
16. Smith, S.W., The role of chelation in the treatment of other metal poisonings. *J. Med. Toxicol.*, 9, 4, 355–369, 2013.
17. Bhardwaj, L.K., Kumar, D., Kumar, A., Phytoremediation potential of Ocimum sanctum: A sustainable approach for remediation of heavy metals, in: *Phytoremediation potential of medicinal and aromatic plants*, pp. 46–57, CRC Press, United Kingdom, 2023.
18. Sarma, H., Metal hyperaccumulation in plants: a review focusing on phytoremediation technology. *J. Environ. Sci. Technol.*, 4, 2, 118–138, 2011.
19. Sakakibara, M., Ohmori, Y., Ha, N.T.H., Sano, S., Sera, K., Phytoremediation of heavy metal-contaminated water and sediment by Eleocharis acicularis. *CLEAN–Soil Air Water*, 39, 8, 735–741, 2011.
20. Miranda Pazcel, E.M., Wannaz, E.D., Pignata, M.L., Salazar, M.J., Tagetes minuta L. variability in terms of lead phytoextraction from polluted soils: is historical exposure a determining factor?. *Environ. Processes*, 5, 2, 243–259, 2018.
21. Gunduz, S., Uygur, F.N., Kahramanoğlu, I., Heavy metal Phytoremediation potentials of Lepidum sativum L., Lactuca sativa L., Spinacia oleracea L. and Raphanus sativus L. *Her. J. Agric. Food Sci.*, 1, 1, 1–5, 2012.
22. Rai, P.K., Phytoremediation of Hg and Cd from industrial effluents using an aquatic free floating macrophyte Azolla pinnata. *Int. J. Phytorem.*, 10, 5, 430–439, 2008.
23. Kumar, V., Kumar, P., Singh, J., Kumar, P., Potential of water fern (Azolla pinnata R., Br.) in phytoremediation of integrated industrial effluent of SIIDCUL, Haridwar, India: removal of physicochemical and heavy metal pollutants. *Int. J. Phytorem.*, 22, 4, 392–403, 2020.
24. Ahmad, R., Tehsin, Z., Malik, S.T., Asad, S.A., Shahzad, M., Bilal, M., Shah, M.M., Khan, S.A., Phytoremediation potential of hemp (*Cannabis sativa* L.): identification and characterization of heavy metals responsive genes. *CLEAN–Soil Air Water*, 44, 2, 195–201, 2016.

25. Koptsik, G.N., Problems and prospects concerning the phytoremediation of heavy metal polluted soils: a review. *Eurasian Soil Sci.*, 47, 9, 923–939, 2014.
26. Boechat, C.L., Carlos, F.S., Nascimento, C.W.A.D., Quadros, P.D.D., Sá, E.L.S.D., Camargo, F.A.D.O., Bioaugmentation-assisted phytoremediation of As, Cd, and Pb using Sorghum bicolor in a contaminated soil of an abandoned gold ore processing plant. *Rev. Bras. Cienc. Solo*, 44, e0200081, 2020.
27. Ahmed, D.A.E.A., Gheda, S.F., Ismail, G.A., Efficacy of two seaweeds dry mass in bioremediation of heavy metal polluted soil and growth of radish (*Raphanus sativus* L.) plant. *Environ. Sci. Pollut. Res.*, 28, 10, 12831–12846, 2021.
28. Sumiahadi, A. and Acar, R., A review of phytoremediation technology: heavy metals uptake by plants. *IOP Conf. Ser.: Earth Environ. Sci.*, 142, 012023, 2018, IOP Publishing.
29. Mishra, V.K. and Tripathi, B.D., Accumulation of chromium and zinc from aqueous solutions using water hyacinth (Eichhornia crassipes). *J. Hazard. Mater.*, 164, 2–3, 1059–1063, 2009.
30. Mohsenzadeh, F. and Mohammadzadeh, R., Phytoremediation ability of the new heavy metal accumulator plants. *Environ. Eng. Geosci.*, 24, 4, 441–450, 2018.
31. Zhong, Y. and Chen, J., Ameliorative effects of Lanthanum (III) on Copper (II) stressed rice (Oryza sativa) and its molecular mechanism revealed by transcriptome profiling. *Plant Physiol. Biochem.*, 152, 184–193, 2020.
32. Kocoń, A. and Jurga, B., The evaluation of growth and phytoextraction potential of Miscanthus x giganteus and Sida hermaphrodita on soil contaminated simultaneously with Cd, Cu, Ni, Pb, and Zn. *Environ. Sci. Pollut. Res.*, 24, 5, 4990–5000, 2017.
33. Sharma, R., Bhardwaj, R., Gautam, V., Bali, S., Kaur, R., Kaur, P., Sharma, M., Kumar, V., Sharma, A., Sonia Thukral, A.K., Phytoremediation in waste management: hyperaccumulation diversity and techniques, in: *Plants under metal and metalloid stress: responses, tolerance and remediation*, pp. 277–302, Springer Singapore, Singapore, 2018.
34. Park, J., Kim, J.Y., Kim, K.W., Phytoremediation of soil contaminated with heavy metals using Brassica napus. *Geosyst. Eng.*, 15, 1, 10–18, 2012.
35. Abhilash, P.C., Tripathi, V., Edrisi, S.A., Dubey, R.K., Bakshi, M., Dubey, P.K., Singh, H.B., Ebbs, S.D., Sustainability of crop production from polluted lands. *Energy Ecol. Environ.*, 1, 1, 54–65, 2016.
36. García Martín, J.F., González Caro, M.D.C., López Barrera, M.D.C., Torres García, M., Barbin, D., Mateos, P.Á., Metal accumulation by Jatropha curcas L. adult plants grown on heavy metal-contaminated soil. *Plants*, 9, 4, 418, 2020.
37. Wang, J., Feng, X., Anderson, C.W., Xing, Y., Shang, L., Remediation of mercury contaminated sites–a review. *J. Hazard. Mater.*, 221, 1–18, 2012.
38. Carrasco-Gil, S., Siebner, H., LeDuc, D.L., Webb, S.M., Millán, R., Andrews, J.C., Hernández, L.E., Mercury localization and speciation in plants grown

hydroponically or in a natural environment. *Environ. Sci. Technol.*, 47, 7, 3082–3090, 2013.
39. Rahman, Z. and Singh, V.P., Bioremediation of toxic heavy metals (THMs) contaminated sites: concepts, applications and challenges. *Environ. Sci. Pollut. Res.*, 27, 22, 27563–27581, 2020.
40. Bani, A., Pavlova, D., Echevarria, G., Mullaj, A., Reeves, R.D., Morel, J.L., Sulçe, S., Nickel hyperaccumulation by the species of Alyssum and Thlaspi (Brassicaceae) from the ultramafic soils of the Balkans. *Bot. Serb.*, 34, 1, 3–14, 2010.
41. Sahay, S., Inam, A., Iqbal, S., Risk analysis by bioaccumulation of Cr, Cu, Ni, Pb and Cd from wastewater-irrigated soil to Brassica species. *Int. J. Environ. Sci. Technol.*, 17, 5, 2889–2906, 2020.
42. Kanwar, V.S., Sharma, A., Srivastav, A.L., Rani, L., Phytoremediation of toxic metals present in soil and water environment: a critical review. *Environ. Sci. Pollut. Res.*, 27, 36, 44835–44860, 2020.
43. Shehata, H.S. and Galal, T.M., Trace metal concentration in planted cucumber (*Cucumis sativus* L.) from contaminated soils and its associated health risks. *J. Consum. Prot. Food Saf.*, 15, 3, 205–217, 2020.
44. Saini, V.K., Suthar, S., Karmveer, C., Kumar, K., Valorization of toxic weed Lantana camara L. biomass for adsorptive removal of lead. *J. Chem.*, 2017, 1, 5612594, 2017.
45. Angelova, V.R., Grekov, D.F., Kisyov, V.K., Ivanov, K.I., Potential of lavender (*Lavandula vera* L.) for phytoremediation of soils contaminated with heavy metals. *Int. J. Agric. Biosyst. Eng.*, 9, 5, 522–529, 2015.
46. Checcucci, A., Bazzicalupo, M., Mengoni, A., Exploiting nitrogen-fixing rhizobial symbionts genetic resources for improving phytoremediation of contaminated soils, in: *Enhancing Cleanup of Environmental Pollutants: Volume 1: Biological Approaches*, pp. 275–288, Springer International Publishing, Cham, 2017.
47. Konkolewska, A., Piechalak, A., Ciszewska, L., Antos-Krzemińska, N., Skrzypczak, T., Hanć, A., Sitko, K., Małkowski, E., Barałkiewicz, D., Małecka, A., Combined use of companion planting and PGPR for the assisted phytoextraction of trace metals (Zn, Pb, Cd). *Environ. Sci. Pollut. Res.*, 27, 12, 13809–13825, 2020.
48. Dinh, N., van Der Ent, A., Mulligan, D.R., Nguyen, A.V., Zinc and lead accumulation characteristics and *in vivo* distribution of Zn2+ in the hyperaccumulator Noccaea caerulescens elucidated with fluorescent probes and laser confocal microscopy. *Environ. Exp. Bot.*, 147, 1–12, 2018.
49. Wan, X., Lei, M., Yang, J., Two potential multi-metal hyperaccumulators found in four mining sites in Hunan Province, China. *Catena*, 148, 67–73, 2017.
50. Asante-Badu, B., Kgorutla, L.E., Li, S.S., Danso, P.O., Xue, Z., Qiang, G., Phytoremediation of organic and inorganic compounds in a natural and an agricultural environment: a review. *Appl. Ecol. Environ. Res.*, 18, 5, 2020.

51. Mitra, A., Chatterjee, S., Voronina, A.V., Walther, C., Gupta, D.K., Lead toxicity in plants: a review, in: *Lead in Plants and the Environment*, pp. 99–116, Springer International Publishing AG, Cham, Switzerland, 2019.
52. Manzoor, M., Gul, I., Manzoor, A., Kamboh, U.R., Hina, K., Kallerhoff, J., Arshad, M., Lead availability and phytoextraction in the rhizosphere of Pelargonium species. *Environ. Sci. Pollut. Res.*, 27, 32, 39753–39762, 2020.
53. Alaboudi, K.A., Ahmed, B., Brodie, G., Phytoremediation of Pb and Cd contaminated soils by using sunflower (Helianthus annuus) plant. *Ann. Agric. Sci.*, 63, 1, 123–127, 2018.
54. Angelova, V.R., Ivanova, R.V., Todorov, G.M., Ivanov, K.I., Potential of Salvia sclarea L. for phytoremediation of soils contaminated with heavy metals. *Int. J. Agric. Biosyst. Eng.*, 10, 780–790, 2016.
55. Srivastava, N., Phytomicrobiome: Synergistic Relationship in Bioremediation of Soil for Sustainable Agriculture, in: *Phytomicrobiome Interactions and Sustainable Agriculture*, pp. 150–163, Wiley Publishers, Hoboken, NJ, USA, 2021.
56. Hu, P.J., Qiu, R.L., Senthilkumar, P., Jiang, D., Chen, Z.W., Tang, Y.T., Liu, F.J., Tolerance, accumulation and distribution of zinc and cadmium in hyperaccumulator Potentilla griffithii. *Environ. Exp. Bot.*, 66, 2, 317–325, 2009.
57. Spielmann, J., Ahmadi, H., Scheepers, M., Weber, M., Nitsche, S., Carnol, M., Bosman, B., Kroymann, J., Motte, P., Clemens, S., Hanikenne, M., The two copies of the zinc and cadmium ZIP6 transporter of Arabidopsis halleri have distinct effects on cadmium tolerance. *Plant Cell Environ.*, 43, 9, 2143–2157, 2020.
58. Ibañéz, S.G., Oller, A.L.W., Paisio, C.E., Alderete, L.G.S., González, P.S., Medina, M.I., Agostini, E., The challenges of remediating metals using phytotechnologies, in: *Heavy Metals in the Environment: Microorganisms and Bioremediation*, pp. 173–191, CRC Press, Boca Raton, FL, USA, 2018.
59. Yadav, K.K., Gupta, N., Kumar, V., Singh, J.K., Bioremediation of heavy metals from contaminated sites using potential species: a review. *Indian J. Environ. Prot.*, 37, 1, 65, 2017.
60. Yan, A., Wang, Y., Tan, S.N., Mohd Yusof, M.L., Ghosh, S., Chen, Z., Phytoremediation: a promising approach for revegetation of heavy metal-polluted land. *Front. Plant Sci.*, 11, 513099, 2020.
61. Chandra, S., Gusain, Y.S., Bhatt, A., Metal hyperaccumulator plants and environmental pollution, in: *Microbial Biotechnology in Environmental Monitoring and Cleanup*, pp. 305–317, IGI Global Scientific Publishing, Hershey, PA, USA, 2018.
62. Okpara-Elom, I.A., Onochie, C.C., Elom, M.O., Ezaka, E., Elom, O., Bioremediation of heavy metal polluted soil using plant growth promoting bacteria: an assessment of response. *Bioremediat. J.*, 28, 1, 34–53, 2024.
63. Akpasi, S.O., Anekwe, I.M.S., Tetteh, E.K., Amune, U.O., Shoyiga, H.O., Mahlangu, T.P., Kiambi, S.L., Mycoremediation as a potentially promising

technology: current status and prospects—a review. *Appl. Sci.*, 13, 8, 4978, 2023.
64. Bhakat, K., Chakraborty, A., Islam, E., Characterization of arsenic oxidation and uranium bioremediation potential of arsenic resistant bacteria isolated from uranium ore. *Environ. Sci. Pollut. Res.*, 26, 13, 12907–12919, 2019.
65. Rahman, Z., Thomas, L., Singh, V.P., Biosorption of heavy metals by a lead (Pb) resistant bacterium, Staphylococcus hominis strain AMB-2. *J. Basic Microbiol.*, 59, 5, 477–486, 2019.
66. Ibarrolaza, A., Coppotelli, B.M., Del Panno, M.T., Donati, E.R., Morelli, I.S., Application of the knowledge-based approach to strain selection for a bioaugmentation process of phenanthrene-and Cr (VI)-contaminated soil. *J. Appl. Microbiol.*, 111, 1, 26–35, 2011.
67. Wang, S., Liu, T., Xiao, X., Luo, S., Advances in microbial remediation for heavy metal treatment: a mini review. *J. Leather Sci. Eng.*, 3, 1, 1, 2021.
68. Upadhyay, N., Vishwakarma, K., Singh, J., Mishra, M., Kumar, V., Rani, R., Mishra, R.K., Chauhan, D.K., Tripathi, D.K., Sharma, S., Tolerance and reduction of chromium (VI) by Bacillus sp. MNU16 isolated from contaminated coal mining soil. *Front. Plant Sci.*, 8, 778, 2017.
69. Puyen, Z.M., Villagrasa, E., Maldonado, J., Diestra, E., Esteve, I., Solé, A., Biosorption of lead and copper by heavy-metal tolerant Micrococcus luteus DE2008. *Bioresour. Technol.*, 126, 233–237, 2012.
70. Hoyle-Gardner, J., Jones, W., Badisa, V.L., Mwashote, B., Ibeanusi, V., Gaines, T., Lowenthal, H., Tucker, L., Lead metal biosorption and isotherms studies by metal-resistant Bacillus strain MRS-2 bacterium. *J. Basic Microbiol.*, 61, 8, 697–708, 2021.
71. Giwa, O.E. and Ibitoye, F.O., Bioremediation of heavy metal in crude oil contaminated soil using isolated Indigenous microorganism cultured with E coli DE3 BL21. *Int. J. Eng. Appl. Sci.*, 4, 6, 257436, 2017.
72. Imron, M.F., Kurniawan, S.B., Soegianto, A., Characterization of mercury-reducing potential bacteria isolated from Keputih non-active sanitary landfill leachate, Surabaya, Indonesia under different saline conditions. *J. Environ. Manage.*, 241, 113–122, 2019.
73. Mahbub, K.R., Krishnan, K., Megharaj, M., Naidu, R., Bioremediation potential of a highly mercury resistant bacterial strain Sphingobium SA2 isolated from contaminated soil. *Chemosphere*, 144, 330–337, 2016.
74. Njoku, K.L., Akinyede, O.R., Obidi, O.F., Microbial remediation of heavy metals contaminated media by Bacillus megaterium and Rhizopus stolonifer. *Sci. Afr.*, 10, e00545, 2020.
75. Mohd, S., Kushwaha, A.S., Shukla, J., Mandrah, K., Shankar, J., Arjaria, N., Saxena, P.N., Khare, P., Narayan, R., Dixit, S., Siddiqui, M.H., Fungal mediated biotransformation reduces toxicity of arsenic to soil dwelling microorganism and plant. *Ecotoxicol. Environ. Saf.*, 176, 108–118, 2019.

76. Sánchez-Castellón, J., Urango-Cárdenas, I., Enamorado-Montes, G., Burgos-Nuñez, S., Marrugo-Negrete, J., Díez, S., Removal of Mercury, Cadmium, and Lead Ions by Penicillium sp. *Front. Environ. Chem.*, 2, 795632, 2022.
77. Herliana, O., Soesanto, L., Mawadah, E., Phytobioremediation of cadmium-contaminated soil using combination of Ipomoea reptans Poir and Trichoderma sp. and its effect on spinach growth and yield. *J. Degraded Min. Lands Manage.*, 6, 1, 1519, 2018.
78. Chang, J., Zhang, H., Cheng, H., Yan, Y., Chang, M., Cao, Y., Huang, F., Zhang, G., Yan, M., Spent Ganoderma lucidum substrate derived biochar as a new bio-adsorbent for Pb2+/Cd2+ removal in water. *Chemosphere*, 241, 125121, 2020.
79. Xu, H., Hao, R.X., Xu, X.Y., Ding, Y., Lu, A.H., Li, Y.H., Removal of hexavalent chromium by Aspergillus niger through reduction and accumulation. *Geomicrobiol. J.*, 38, 1, 20–28, 2021.
80. Dusengemungu, L., Kasali, G., Gwanama, C., Ouma, K.O., Recent advances in biosorption of copper and cobalt by filamentous fungi. *Front. Microbiol.*, 11, 582016, 2020.
81. Al-Hagar, O.E., Bayoumi, R.A., Abdel Aziz, O.A., Mousa, A.M., Biosorption and adsorption of some heavy metals by Fusarium sp. F6c isolate as affected by gamma irradiation and agricultural wastes. *ScienceAsia*, 46, 1, 2020.
82. Chatterjee, A., Das, R., Abraham, J., Bioleaching of heavy metals from spent batteries using Aspergillus nomius JAMK1. *Int. J. Environ. Sci. Technol.*, 17, 1, 49–66, 2020.
83. Infante, J.C., De Arco, R.D., Angulo, M.E., Removal of lead, mercury and nickel using the yeast Saccharomyces cerevisiae. *Rev. MVZ Córdoba*, 19, 2, 4141–4149, 2014.
84. Senthilkumar, R., Reddy Prasad, D.M., Govindarajan, L., Saravanakumar, K., Naveen Prasad, B.S., Synthesis of green marine algal-based biochar for remediation of arsenic (V) from contaminated waters in batch and column mode of operation. *Int. J. Phytorem.*, 22, 3, 279–286, 2020.
85. Kumar, K.S., Dahms, H.U., Won, E.J., Lee, J.S., Shin, K.H., Microalgae–a promising tool for heavy metal remediation. *Ecotoxicol. Environ. Saf.*, 113, 329–352, 2015.
86. Verma, N. and Sharma, R., Bioremediation of toxic heavy metals: a patent review. *Recent Pat. Biotechnol.*, 11, 3, 171–187, 2017.
87. Shanab, S., Essa, A., Shalaby, E., Bioremoval capacity of three heavy metals by some microalgae species (Egyptian Isolates). *Plant Signal. Behav.*, 7, 3, 392–399, 2012.
88. Al-Homaidan, A.A., Al-Qahtani, H.S., Al-Ghanayem, A.A., Ameen, F., Ibraheem, I.B., Potential use of green algae as a biosorbent for hexavalent chromium removal from aqueous solutions. *Saudi J. Biol. Sci.*, 25, 8, 1733–1738, 2018.

89. Hamouda, R.A., El-Naggar, N.E.A., Doleib, N.M., Saddiq, A.A., Bioprocessing strategies for cost-effective simultaneous removal of chromium and malachite green by marine alga Enteromorpha intestinalis. *Sci. Rep.*, 10, 1, 13479, 2020.
90. Borah, D., Kennedy, B., Gopalakrishnan, S., Chithonirai, A., Nooruddin, T., Bioremediation and biomass production with the green microalga Chlorococcum humicola and textile mill effluent (TE). *Proc. Natl. Acad. Sci., India, Sect. B Biol. Sci.*, 90, 2, 415–423, 2020.
91. Saavedra, R., Muñoz, R., Taboada, M.E., Vega, M., Bolado, S., Comparative uptake study of arsenic, boron, copper, manganese and zinc from water by different green microalgae. *Bioresour. Technol.*, 263, 49–57, 2018.
92. Buayam, N., Davey, M.P., Smith, A.G., Pumas, C., Effects of Copper and pH on the Growth and Physiology of Desmodesmus sp. AARLG074. *Metabolites*, 9, 5, 84, 2019.
93. Deng, L., Su, Y., Su, H., Wang, X., Zhu, X., Sorption and desorption of lead (II) from wastewater by green algae Cladophora fascicularis. *J. Hazard. Mater.*, 143, 1–2, 220–225, 2007.
94. He, H.J., Xiang, Z.H., Chen, X.J., Chen, H., Huang, H., Wen, M., Yang, C.P., Biosorption of Cd (II) from synthetic wastewater using dry biofilms from biotrickling filters. *Int. J. Environ. Sci. Technol.*, 15, 7, 1491–1500, 2018.
95. Holan, Z.R. and Volesky, B., Biosorption of lead and nickel by biomass of marine algae. *Biotechnol. Bioeng.*, 43, 11, 1001–1009, 1994.
96. López-Miranda, J.L., Silva, R., Molina, G.A., Esparza, R., Hernandez-Martinez, A.R., Hernández-Carteño, J., Estévez, M., Evaluation of a dynamic bioremediation system for the removal of metal ions and toxic dyes using Sargassum Spp. *J. Mar. Sci. Eng.*, 8, 11, 899, 2020.
97. Henriques, B., Rocha, L.S., Lopes, C.B., Figueira, P., Monteiro, R.J., Duarte, A.D.C., Pardal, M.A., Pereira, E., Study on bioaccumulation and biosorption of mercury by living marine macroalgae: prospecting for a new remediation biotechnology applied to saline waters. *Chem. Eng. J.*, 281, 759–770, 2015.
98. Shamshad, I., Khan, S., Waqas, M., Asma, M., Nawab, J., Gul, N., Raiz, A., Li, G., Heavy metal uptake capacity of fresh water algae (Oedogonium westti) from aqueous solution: A mesocosm research. *Int. J. Phytorem.*, 18, 4, 393–398, 2016.
99. Jin, Y., Luan, Y., Ning, Y., Wang, L., Effects and mechanisms of microbial remediation of heavy metals in soil: a critical review. *Appl. Sci.*, 8, 8, 1336, 2018.
100. Sun, X., Liu, Z., Jiang, Q., Yang, Y., Concentrations of various elements in seaweed and seawater from Shen'ao Bay, Nan'ao Island, Guangdong coast, China: Environmental monitoring and the bioremediation potential of the seaweed. *Sci. Total Environ.*, 659, 632–639, 2019.
101. Bahafid, W., Tahri Joutey, N., Sayel, H., Boularab, I., El Ghachtouli, N., Bioaugmentation of chromium-polluted soil microcosms with Candida

tropicalis diminishes phytoavailable chromium. *J. Appl. Microbiol.*, 115, 3, 727–734, 2013.

102. Benazir, J.F., Suganthi, R., Rajvel, D., Pooja, M.P., Mathithumilan, B., Bioremediation of chromium in tannery effluent by microbial consortia. *Afr. J. Biotechnol.*, 9, 21, 3140–3143, 2010.
103. El-Naggar, N.E.A., Hamouda, R.A., Mousa, I.E., Abdel-Hamid, M.S., Rabei, N.H., Biosorption optimization, characterization, immobilization and application of Gelidium amansii biomass for complete Pb2+ removal from aqueous solutions. *Sci. Rep.*, 8, 1, 13456, 2018.
104. De, J., Ramaiah, N., Vardanyan, L., Detoxification of toxic heavy metals by marine bacteria highly resistant to mercury. *Mar. Biotechnol.*, 10, 4, 471–477, 2008.
105. Hassan, A., Pariatamby, A., Ahmed, A., Auta, H.S., Hamid, F.S., Enhanced bioremediation of heavy metal contaminated landfill soil using filamentous fungi consortia: a demonstration of bioaugmentation potential. *Water Air Soil Pollut.*, 230, 9, 215, 2019.
106. Polti, M.A., Aparicio, J.D., Benimeli, C.S., Amoroso, M.J., Simultaneous bioremediation of Cr (VI) and lindane in soil by actinobacteria. *Int. Biodeterior. Biodegrad.*, 88, 48–55, 2014.
107. Pan, R., Cao, L., Zhang, R., Combined effects of Cu, Cd, Pb, and Zn on the growth and uptake of consortium of Cu-resistant Penicillium sp. A1 and Cd-resistant Fusarium sp. A19. *J. Hazard. Mater.*, 171, 1–3, 761–766, 2009.
108. Nwaehiri, U.L., Akwukwaegbu, P.I., Nwoke, B.E.B., Bacterial remediation of heavy metal polluted soil and effluent from paper mill industry. *Environ. Anal. Health Toxicol.*, 35, 2, e2020009, 2020.
109. Prashant and Kumar, S., Technological Outline of Constructed Wetlands: An Alternative for Sustainable and Decentralized Wastewater Treatment, in: *Recent Trends in Wastewater Treatment*, pp. 51–74, Springer International Publishing, Cham, 2022.
110. Gruiz, K., Natural attenuation in contaminated soil remediation, in: *Engineering Tools for Environmental Risk Management*, vol. 4, pp. 95–201, Taylor & Francis Group, London, United Kingdom, 2019.
111. Bayar, J., Ali, N., Dong, Y., Ahmad, U., Anjum, M.M., Khan, G.R., Zaib, M., Jalal, A., Ali, R., Ali, L., Biochar-based adsorption for heavy metal removal in water: a sustainable and cost-effective approach. *Environ. Geochem. Health*, 46, 11, 428, 2024.
112. Motlagh, A., Urban Stormwater and Groundwater Quality: Pathways, Risks, and Green Infrastructure Solutions. *Environments*, 12, 11, 446, 2025.
113. Kumar, S., Aspects of Genetically Modified Plants in Removing Heavy Metals From the Soil, in: *Policy Issues in Genetically Modified Crops*, pp. 273–289, Academic Press, San Diego, CA, USA, 2021.
114. Valls, M. and De Lorenzo, V., Exploiting the genetic and biochemical capacities of bacteria for the remediation of heavy metal pollution. *FEMS Microbiol. Rev.*, 26, 4, 327–338, 2002.

115. Verma, S., Verma, P.K., Chakrabarty, D., Arsenic bio-volatilization by engineered yeast promotes rice growth and reduces arsenic accumulation in grains. *Int. J. Environ. Res.*, 13, 3, 475–485, 2019.
116. Liu, S., Zhang, F., Chen, J., Sun, G., Arsenic removal from contaminated soil *via* biovolatilization by genetically engineered bacteria under laboratory conditions. *J. Environ. Sci.*, 23, 9, 1544–1550, 2011.
117. Singh, S., Kang, S.H., Mulchandani, A., Chen, W., Bioremediation: environmental clean-up through pathway engineering. *Curr. Opin. Biotechnol.*, 19, 5, 437–444, 2008.
118. Ivask, A., Dubourguier, H.C., Põllumaa, L., Kahru, A., Bioavailability of Cd in 110 polluted topsoils to recombinant bioluminescent sensor bacteria: effect of soil particulate matter. *J. Soils Sediments*, 11, 2, 231–237, 2011.
119. Patel, J., Zhang, Q., McKay, R.M.L., Vincent, R., Xu, Z., Genetic engineering of Caulobacter crescentus for removal of cadmium from water. *Appl. Biochem. Biotechnol.*, 160, 1, 232–243, 2010.
120. Wu, C.H., Wood, T.K., Mulchandani, A., Chen, W., Engineering plant-microbe symbiosis for rhizoremediation of heavy metals. *Appl. Environ. Microbiol.*, 72, 2, 1129–1134, 2006.
121. Sriprang, R., Hayashi, M., Ono, H., Takagi, M., Hirata, K., Murooka, Y., Enhanced accumulation of Cd2+ by a Mesorhizobium sp. transformed with a gene from Arabidopsis thaliana coding for phytochelatin synthase. *Appl. Environ. Microbiol.*, 69, 3, 1791–1796, 2003.
122. Ackerley, D.F., Gonzalez, C.F., Keyhan, M., Blake, R., Matin, A.J.E.M., Mechanism of chromate reduction by the Escherichia coli protein, NfsA, and the role of different chromate reductases in minimizing oxidative stress during chromate reduction. *Environ. Microbiol.*, 6, 8, 851–860, 2004.
123. Al Hasin, A., Gurman, S.J., Murphy, L.M., Perry, A., Smith, T.J., Gardiner, P.H., Remediation of chromium (VI) by a methane-oxidizing bacterium. *Environ. Sci. Technol.*, 44, 1, 400–405, 2010.
124. Bondarenko, O., Rõlova, T., Kahru, A., Ivask, A., Bioavailability of Cd, Zn and Hg in soil to nine recombinant luminescent metal sensor bacteria. *Sensors*, 8, 11, 6899–6923, 2008.
125. Sasaki, Y., Minakawa, T., Miyazaki, A., Silver, S., Kusano, T., Functional dissection of a mercuric ion transporter, MerC, from Acidithiobacillus ferrooxidans. *Biosci. Biotechnol. Biochem.*, 69, 7, 1394–1402, 2005.
126. Kiyono, M., Sone, Y., Nakamura, R., Pan-Hou, H., Sakabe, K., The MerE protein encoded by transposon Tn21 is a broad mercury transporter in Escherichia coli. *FEBS Lett.*, 583, 7, 1127–1131, 2009.
127. Ng, S.P., Davis, B., Palombo, E.A., Bhave, M., A Tn 5051-like mer-containing transposon identified in a heavy metal tolerant strain Achromobacter sp. AO22. *BMC Res. Notes*, 2, 1, 38, 2009.
128. Bae, W., Wu, C.H., Kostal, J., Mulchandani, A., Chen, W., Enhanced mercury biosorption by bacterial cells with surface-displayed MerR. *Appl. Environ. Microbiol.*, 69, 6, 3176–3180, 2003.

129. Brim, H., McFarlan, S.C., Fredrickson, J.K., Minton, K.W., Zhai, M., Wackett, L.P., Daly, M.J., Engineering Deinococcus radiodurans for metal remediation in radioactive mixed waste environments. *Nat. Biotechnol.*, 18, 1, 85–90, 2000.
130. Ruiz, O.N., Alvarez, D., Gonzalez-Ruiz, G., Torres, C., Characterization of mercury bioremediation by transgenic bacteria expressing metallothionein and polyphosphate kinase. *BMC Biotechnol.*, 11, 1, 82, 2011.
131. Deng, X., Yi, X.E., Liu, G., Cadmium removal from aqueous solution by gene-modified Escherichia coli JM109. *J. Hazard. Mater.*, 139, 2, 340–344, 2007.
132. Freeman, J.L., Persans, M.W., Nieman, K., Salt, D.E., Nickel and cobalt resistance engineered in Escherichia coli by overexpression of serine acetyltransferase from the nickel hyperaccumulator plant Thlaspi goesingense. *Appl. Environ. Microbiol.*, 71, 12, 8627–8633, 2005.
133. Singh, A., Tyagi, P., Ranjan, R., Sushkova, S.N., Minkina, T., Burachevskaya, M., Rajput, V.D., Bioremediation of hazardous wastes using green synthesis of nanoparticles. *Processes*, 11, 1, 141, 2023.
134. Amin, I., Nazir, R., Rather, M.A., Nano-bioremediation: an innovative approach for remedying heavy metals using fungi. *J. Bioremediat. Biodegrad.*, 12, 487, 2, 2021.
135. Mahour, S. and Srivastava, S., Biochar-Based nanocomposites for separation of inorganic contaminants from the environment, in: *Biochar-Based Nanocomposites for Contaminant Management: Synthesis, Contaminants Removal, and Environmental Sustainability*, pp. 69–81, Springer International Publishing, Cham, 2023.

7

Rhizobacteria Can Act as a Sustainable Asset to Mitigate Heavy Metal and Organic Pollution Stress and Induce Plant Growth

Sakshi Tomar[1], Induja Mishra[2], Pashupati Nath[2], Bhupendra Singh[1] and Prabhat K. Chauhan[1]*

[1]*Department of Environmental Science, AKS University, Satna, Madhya Pradesh, India*
[2]*School of Applied and Life Sciences & UIT, Uttaranchal University, Dehradun, Uttarakhand, India*

Abstract
The accelerating pace of industrialization, urbanization, and intensive farming has triggered a global soil contamination crisis, characterized by rising levels of toxic heavy metals (HMs) and persistent organic pollutants (POPs). These pollutants not only degrade soil health but also infiltrate the food chain, posing serious risks to environmental sustainability and human well-being. In the context of achieving the United Nations Sustainable Development Goals (UNSDGs) by 2030, there is an urgent need for innovative and sustainable remediation strategies. This chapter explores the potential of Plant Growth-Promoting Rhizobacteria (PGPR) as a groundbreaking, eco-friendly solution that enhances crop productivity while remediating contaminated soils. By harnessing the natural biocontrol and detoxification capabilities of PGPR, we can effectively reduce the harmful impacts of HMs and POPs, strengthen food security, and restore ecological balance. These microbial allies represent a promising path toward resilient agriculture and a healthier, more sustainable planet.

Keywords: Emerging contaminants, heavy metals, organic pollutants, phytostimulation, plant growth-promoting rhizobacteria, soil restoration

**Corresponding author*: cprabhatenv@gmail.com; ORCID: https://orcid.org/0009-0000-5902-1613

Prangya Rath, Laxmi Kant Bhardwaj and Amit Kumar (eds.) Heavy Metal Remediation, (171–194)

7.1 Introduction

Food security is fundamental to human health and well-being, yet ensuring food safety throughout the entire supply chain from production to consumption remains a significant challenge. Environmental pollution, fueled by rapid industrialization, urban expansion, and intensive agricultural practices, poses a growing danger to global food security. Achieving the Sustainable Development Goals (SDGs) requires not only boosting crop productivity but also safeguarding food quality. The infiltration of unexpected pollutants, including toxic HMs, POPs, and emerging contaminants (ECs) such as pharmaceuticals, microplastics (MPs), and phenols into the agricultural system presents serious risks to both human health and ecosystems [1, 2]. HMs like arsenic (As), cadmium (Cd), and lead (Pb), primarily released through industrial and urban activities, are of particular concern in densely populated nations such as India and China [3]. Additionally, the widespread occurrence of POPs (e.g., dichlorodiphenyltrichloroethane (DDT), polychlorinated biphenyls (PCBs)) and unregulated ECs (e.g., ibuprofen, caffeine, steroidal estrogens) continues to contaminate natural ecosystems [4]. Their increasing prevalence highlights the urgent need for sustainable, science-driven interventions to protect the food system and ensure public health (Figure 7.1).

Pollutant concentrations in irrigated agricultural water typically range from 10.0 to 5000.0 ng/L, with corresponding levels in crops reaching up to 7500.0 ng/kg [5]. It is estimated that up to 10.0 g of these contaminants may enter the human body daily through food consumption. These toxic

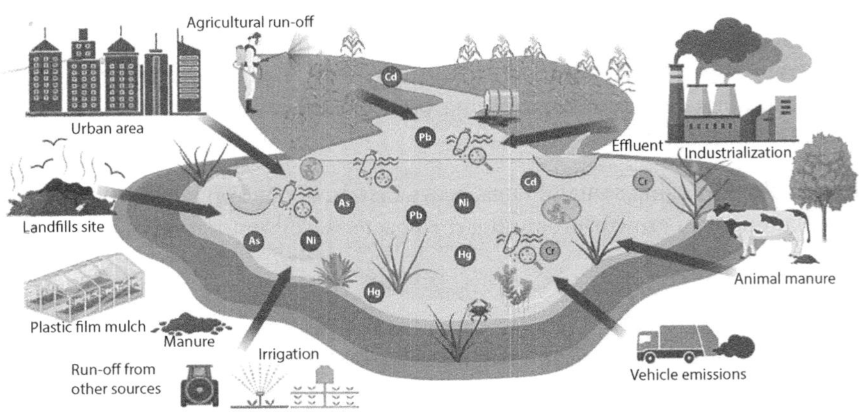

Figure 7.1 Heavy metal (HM) and emerging contaminants (ECs) pollution from different environmental sources.

substances interfere with vital plant functions such as photosynthesis, nutrient uptake, and hormone regulation, ultimately leading to reduced growth, biomass, and reproductive capacity. To combat the adverse effects of HMs and ECs, researchers have explored a variety of biotechnological strategies, including rhizoremediation, mycoremediation, biostimulation, cyanoremediation, biomineralization, phytostabilization, biosorption, hyper-accumulation, dendro-remediation, and geno-remediation. Despite their promise, concerns over biosafety, ecological risks, and limited field applicability remain major challenges. Innovative tools such as nanotechnology (NT) and clustered regularly interspaced short palindromic repeats (CRISPR) also hold significant potential for remediation, but their environmental and health implications require thorough evaluation before large-scale implementation.

PGPR plays a crucial role in alleviating contaminant-induced stress and enhancing crop performance. Numerous studies have demonstrated the effectiveness of microbial consortia, such as *Pseudomonas putida, Bacillus subtilis, Rhizobium-Azospirillum brasilense, Pseudomonas fluorescens,* and *Arthrobacter species,* in promoting plant resilience under exposure to HMs and ECs [6].

Plant Growth-Promoting Rhizobacteria (PGPR) promote plant growth through phytostimulation, biofertilizer, and biocontrol activities extensively reported by several workers across the globe, while PGPR is also reported for the mitigation of toxic HM by transformation, adsorption, bioaccumulation, and other activities, but very few reports are available that reflect PGPR degraded organic pollutants/emerging pollutants. Although rhizobacteria can release different enzymes such as reductase, oxidases, etc., can degrade POP therefore, keeping these views the present study demonstrates with the aim (i) the Brutal impact of toxic HM and ECs on crop production, (ii) Systematic knowledge on PGPR mediated toxic HM removal and degradation of organic component, (iii) Development of possible sustainable management strategy mediated PGPR to mitigate contaminated aforesaid issues for plant growth.

7.2 The Brutal Impact of Heavy Metal (HM) and Emerging Contaminant (EC) on Crop Production

Heavy metals (HMs) and EC enter agricultural soils from the agricultural practices, industrial emissions, livestock manure, wastewater irrigation, metal-based pesticides, phosphate fertilizers, and sewage sludge [7].

HMs and ECs, toxic even at low levels, induce stress responses like chlorosis, reactive oxygen species (ROS) generation, enzyme disruption, ionic imbalance, and deoxyribonucleic acid (DNA) damage, while their accumulation in edible crop parts such as As in rice from contaminated groundwater poses serious health risks [8], Vanadium (V) stress in sweet potato significantly reduces leaf area, biomass, and chlorophyll content [9] while also inhibiting germination and root growth in soybeans [10]. Chromium (Cr) exposure decreases chlorophyll and nitrogen (N) levels, impairs plant growth, and induces oxidative stress [11, 12]. Similarly, Cd reduces enzymatic activity and increases oxidative damage in spinach and rice [13, 14]. Vanadium (V) stress decreased leaf area by 7.6%, 11.7% reduction in leaf count, shoot fresh weight (14%), root fresh weight (16%), shoot dry weight (25.5%), root dry weight (28.9%) and chlorophyll content by 77% and carotenoid levels by 50.3% in *Ipomoea batatas* [9]. Excess As and Cd disrupt enzyme function, causing oxidative damage and visible plant injury [13, 15]. HMs also affect soil microbial communities and enzyme activity, disrupting nutrient cycling and organic matter decomposition [16]. Leaching of HMs into groundwater can threaten aquatic ecosystems and human health [3, 17, 18]. Additionally, ECs from anthropogenic sources such as pharmaceuticals and personal care products (PCPs) are increasingly detected in agricultural soils and pose new ecological and health risks [5] (Table 7.1).

Therefore, the mitigation of HMs and ECs in agriculture is imperative to safeguard food security, nutritional quality, and soil health in alignment with the SDGs 2030 agenda.

7.3 Mechanisms and Functions of Plant Growth-Promoting Rhizobacteria (PGPR)

Plant Growth-Promoting Rhizobacteria (PGPR), a vital group of root-associated microbes, enhance plant development through a range of direct and indirect mechanisms, including phytostimulation, mineralization, nutrient solubilization, pollutant degradation, absorption, and biocontrol [6]. PGPR plays a key role in mitigating HM toxicity by immobilizing metals through biosorption, bioaccumulation, and precipitation processes [27]. Their high surface area, extracellular polymeric substance, and metal-binding ligands contribute to effective adsorption and detoxification of HMs, thereby facilitating the removal of metals from contaminated soils and water [27]. Moreover, PGPR enhances plant tolerance to

Table 7.1 Effects of heavy metals (HMs) on crops and plant physiology.

S. no.	Heavy metals (HMs)	Crops\plants	Effects	References
1	Arsenic (As)	*Oryza sativa*	Reduced plant growth, chlorophyll content, antioxidants, and enzymatic activities, and increased oxidative stress.	[15]
		Brassica juncea, B. oleracea, Raphanus sativus,	High concentration of arsenic (As) reduces the shoot and root growth of the seedling.	[19]
		Glycine max L.	Reduces plant growth, root absorption rate, water and osmotic potential, and xylem vessels.	[20]
2	Cadmium (Cd)	*Spinacia oleracea*	The high concentration of cadmium (Cd) affects the alteration of the enzymatic activities and lowers the catabolic proteins.	[13]
		Oryza sativa	Reduced plant growth, chlorophyll content, and photosynthetic rate increased oxidative damage.	[14]
		Brassica juncea	High concentration of cadmium (Cd) led to reduced plant growth, biomass, and relative water content, and chlorophyll content, attributable to heightened oxidative damage.	[21]

(*Continued*)

Table 7.1 Effects of heavy metals (HMs) on crops and plant physiology. (*Continued*)

S. no.	Heavy metals (HMs)	Crops\plants	Effects	References
3	Chromium (Cr)	*Triticum aestivum*	Reduced plant growth, all other physiological processes, and oxidative damage.	[22]
		Oryza sativa	High concentration of chromium (Cr) leads to reduced plant growth, biomass chlorophyll contents, and overall yields of crops, as well as inducing oxidative stress.	[12]
		Helianthus annuus	Chromium (Cr) stress reduces the photosynthetic rate, transpiration rate, and stomatal conductance.	[23]
		Brassica juncea	Chromium (Cr) toxicity reduces chlorophyll content and carotenoid by 11.05% and 22.93% as well as 37.56% of plant nitrogen (N), as compared to the control.	[11]
4	Lead (Pb)	*Ricinus communis* L	Lead (Pb) stress reduces the plant's growth, photosynthetic rate, and nutrient mineralization.	[24]
5	Lead (Pb) and Copper (Cu)	*Medicago sativa, Triticum aestivum*	The toxicity of lead (Pb^{2+}) and copper (Cu^{2+}) ions suppresses the seed germination, shoot and root length, and overall growth index of plants.	[25]

(*Continued*)

Table 7.1 Effects of heavy metals (HMs) on crops and plant physiology. (*Continued*)

S. no.	Heavy metals (HMs)	Crops\plants	Effects	References
6	Copper (Cu) and Nickel (Ni)	*Triticum aestivum*	High concentration of copper (Cu) decreases yields by about 10% and increasing concentration of copper (Cu) and nickel (Ni) affects the crop yields and wheat straw.	[26]
7	Vanadium (V)	*Glycine max L.*	Inhibited seed germination, biomass production, and reduction of lateral root formation.	[10]
		Ipomoea batatas	Stress decreased 7.6% in leaf area, 11.7% reduction in leaf count, shoot fresh weight (14%), root fresh weight (16%), shoot dry weight (25.5%), root dry weight (28.9%), and chlorophyll content by 77% and carotenoid levels by 50.3%.	[9]

environmental stressors, including toxic HMs and organic pollutants, by modulating physiological and biochemical responses. This study provides an in-depth exploration of the mechanisms and functional roles of PGPR in supporting plant growth and remediating contaminated environments.

7.3.1 Removal of HMs and Plant Growth Promotion

Plant Growth-Promoting Rhizobacteria (PGPR) perform a dual purpose: they improve plant growth and facilitate the removal of HMs from the contaminated environment. PGPR achieve this by releasing organic composites such as siderophores, organic acids, and exopolysaccharides into the rhizosphere, which chelate HMs and reduce their bioavailability to plants [27, 28]. Numerous studies have demonstrated the effectiveness of specific PGPR strains in promoting plant growth while remediating HM-contaminated soils. For example, *Mesorhizobium* RC3 significantly reduced soil Cr levels while improving plant attributes such as shoot and root length, nodulation, chlorophyll content, and yield over a 90-135 day period [29]. Similarly, *Bacillus cereus* enhanced seed germination and biomass in *Brassica nigra,* while *Bacillus* sp. ZC3-2-1 reduced Cd and zinc (Zn) bioavailability, promoting metal uptake in *Oryza sativa* [30, 31] (Table 7.2).

A microbial consortium comprising *Sphingomonas, Nitrospira, Massilia,* and members of the *Xanthobacteraceae* family successfully removed over 86% of di(2-ethylhexyl) phthalate (DEHP), 76% of Cd, and 92% of Zn from contaminated soils, while simultaneously enhancing rice growth [31]. Likewise, the inoculations of *Zea mays* with *B. cereus* TCU11 in soils contaminated with Pb, Zn, Cu, Cd, and Ni significantly improved plant biomass, chlorophyll, carotenoid, and protein content through siderophore production compared to uninoculated controls [32]. These findings underscore the remarkable potential of PGPR and microbial consortia in supporting sustainable phytoremediation while boosting crop productivity in contaminated environments. Additionally, PGPR enhances micronutrient uptake *via* mechanisms such as phytoextraction and phyto-stability, with co-inoculation strategies showing improved Zn translocation in wheat [33]. To further enhance bioremediation efficiency, additional microbial strategies, including biosorption, biotransformation, bioaugmentation, biostimulation, and bioaccumulation, are being increasingly employed [1, 34].

7.3.2 For the Removal of Persistent Organic Pollutants (POPs) and Plant Growth

Plant Growth-Promoting Rhizobacteria (PGPR) and *Rhizobium* species play crucial roles in soil bioremediation by enhancing plant growth and degrading organic contaminants through diverse metabolic pathways [42, 43]. Microbial genera such as *Pseudomonas, Bacillus,* and *Arthrobacter* are particularly effective in breaking down a wide range of organic pollutants while simultaneously supporting plant health through N fixation and nutrient solubilization [44–46] (Table 7.3). For example, *Staphylococcus sp.* and *Bacillus circulans* demonstrated the ability to degrade up to 75.88% of endosulfan while enhancing seeding biomass [47]. Similarly, strains such as *Pseudomonas* and *Serratia* are known to degrade organophosphates and

Table 7.2 Rhizobacteria-mediated remediation of heavy metals (HMs) and plant growth performance in different crops.

S. no.	Rhizobacteria	Remediation of heavy metals (HMs)	Target crops	Plant growth performance	References
1	*Klebsiella variicola* and *Pseudomonas otitidis*	Nickel (Ni) and Lead (Pb)	*Brassica juncea*	The microbial consortia (*Klebsiella variicola and Pseudomonas otitidis*) remediate nickel (Ni) and lead (Pb and increase the vegetative growth in *Brassica juncea* under contaminated soil with 96.05 mg/kg nickel chloride ($NiCl_2$) and 89.63 mg/kg lead nitrate $Pb(NO_3)_2$ compared to the control.	[35]
2	*Bacillus glyciniifermentans*	Zinc (Zn), Chromium (Cr), Copper (Cu), and Nickel (Ni)	*Hordeum vulgare*	Inoculating *Hordeum vulgare* L. with *Bacillus glyciniifermentans* IS-2T under heavy metal (HM) stress improves growth parameters and facilitates the remediation of zinc (Zn), chromium (Cr), copper (Cu), and nickel (Ni).	[36]

(Continued)

Table 7.2 Rhizobacteria-mediated remediation of heavy metals (HMs) and plant growth performance in different crops. (*Continued*)

S. no.	Rhizobacteria	Remediation of heavy metals (HMs)	Target crops	Plant growth performance	References
3	*Bacillus anthracis*	Chromium (Cr) and Cadmium (Cd)	*Sesbania sesban*	Inoculation of *Sesbania sesban* seedlings with *B. anthracis* PM21 significantly boosted seed germination by 5%, root length by 31%, shoot length by 23% photosynthetic pigments such as chlorophyll a by 20%, chlorophyll b by 16% and total chlorophyll by 18% as compared to the control.	[37]
4	*Bacillus cereus*	Chromium (Cr)	*Brassica nigra*	The inoculation of *Brassica nigra* L. with *B. cereus* led to a notable increase in the seed germination by 28.07%, shoot length by 35.86%, root length by 19.11%, fresh biomass by 48.00% and dry biomass by 62.16% compared to the control.	[30]

(*Continued*)

Table 7.2 Rhizobacteria-mediated remediation of heavy metals (HMs) and plant growth performance in different crops. (*Continued*)

S. no.	Rhizobacteria	Remediation of heavy metals (HMs)	Target crops	Plant growth performance	References
5	*Mesorhizobium strain RC3*		Chickpea	The *Mesorhizobium strain RC3* reduced chromium (Cr) content (60.52 mg/Kg) in the soil and enhanced root length by 10.87, shoot length by 21.60%, nodule count by 6.64% and nodule dry weight by 13.77% at 90 days. 135 days and the highest growth in crop seed yield by 27.45%. By 135 days, it further improved root length by 18.05%, shoot length by 21.60%, chlorophyll content by 6.83%, and leghemoglobin content by 9.47%, crop protein content by 16.83% and crop yield by 27.45%.	[29]
6	*Providencia rettgeri, Bacillus cereus,* and *Myroides odoratimimus*		*Sorghum bicolour*	Inoculation of *Sorghum bicolor* with *B. cereus* TCR17, *Providencia rettgeri* TCR21, and *Myroides odoratimimus* TCR22 boosted plant growth and antioxidant status while reducing proline and malondialdehyde contents under chromium (Cr) stress, compared to uninoculated plants.	[38]

(*Continued*)

Table 7.2 Rhizobacteria-mediated remediation of heavy metals (HMs) and plant growth performance in different crops. *(Continued)*

S. no.	Rhizobacteria	Remediation of heavy metals (HMs)	Target crops	Plant growth performance	References
7	*Bacillus cereus*	Lead (Pb), Zinc (Zn), Copper (Cu), Nickel (Ni), and Cadmium (Cd)	*Zea mays*	The treatment of *Zea mays* with *B. cereus* TCU11 in soil contaminated with lead (Pb), zinc (Zn), copper (Cu), cadmium (Cd), and nickel (Ni) resulted in *in-vitro* plant growth promotion through siderophore production, leading to increased plant biomass, chlorophyll, carotenoids, and protein contents compared to the control.	[32]
8	*Sphingomonas*, *Nitrospira*, *Massilia*, and *Xanthobacteraceae*	Copper (Cu) and Zinc (Zn)	*Oryza sativa*	The bacterial consortium efficiently dissipated 86.1% of di(2-ethylhexyl) phthalate (DEHP), removed 76.0% of cadmium (Cd^{2+}), and 92.2% of zinc (Zn^{2+}) from the soil, while also promoting the growth of *Oryza sativa*.	[39]

(Continued)

Table 7.2 Rhizobacteria-mediated remediation of heavy metals (HMs) and plant growth performance in different crops. (*Continued*)

S. no.	Rhizobacteria	Remediation of heavy metals (HMs)	Target crops	Plant growth performance	References
9	*Bacillus sp* ZC3-2-1	Cadmium (Cd) and Zinc (Zn)		The *Bacillus* sp. ZC3-2-1 strain reduced the bioavailable concentrations of cadmium (Cd) and zinc (Zn) in soil by 39.3% and 32.0%, respectively. It enhanced the phytoextraction of cadmium (Cd^{2+}) and zinc (Zn^{2+}) by *Oryza sativa* to 48.2% and 8.0%, respectively, while also significantly increasing plant biomass.	[31]
10	*Pseudomonas fluorescens* and *Trichoderma* spp.	Cadmium (Cd)	Chickpea	The combined inoculation of (*P. fluorescence and Trichoderma* sp. T-4) with chickpea caused maximum increase in germination by 10%, root dry biomass by 71.4%, chlorophyll a by 38% and chlorophyll b by 41%.	[40]
11	*Trichoderma* spp.	Cadmium (Cd) and Lead (Pb)	*Vigna radiata*	*Trichoderma* TF-13 inoculation increased the growth of *Vigna radiata* L., improved germination rate by 10%, root length by 26%, root biomass by 32% and also chlorophyll content.	[41]

(*Continued*)

Table 7.2 Rhizobacteria-mediated remediation of heavy metals (HMs) and plant growth performance in different crops. *(Continued)*

S. no.	Rhizobacteria	Remediation of heavy metals (HMs)	Target crops	Plant growth performance	References
12	*Staphylococcus aureus*	Chromium (Cr)	*Triticum aestivum*	Inoculating *Triticum aestivum* with *S. aureus* strain K1 resulted in a significant enhancement in shoot length (20.6%), root length (27.4%), shoot fresh weight (136%), and shoot dry weight (38.9%) compared to the control.	[15]
			Oryza sativa	Bacterial inoculation elevated the chlorophyll content of *Oryza sativa* from 4.4 to 5.4. Additionally, it raised the photosynthetic rate (from 6.5 to 8.7 µmol H_2O/m^2S^1) and stomatal conductance (from 1.4 to 2.7 µmol H_2O/m^2S^1).	[12]

Table 7.3 Different microbial species involved in organic pollutants degradation and their removal pathway.

S. no.	Degrading microbes	Target organic pollutants	Removal pathways	References
1	*Pseudomonas aeruginosa*	Azo dye	Enzymatic reduction and degradation of the azo bond	[49]
2	*Pseudomonas putida*	Benzene, toluene, and xylene	Aerobic metabolism and bioabsorption	[44]
3	*Pseudomonas putida,* Strain Ch2	Caprolactam and glyphosate	Glyphosate oxidoreductase	[50]
4	*Pseudomonas denitrificans*	Nitrates	Aerobic denitrification process	[53]
5	*Bacillus* sp.	Fenpropathrin	Bio-augmentation	[48]
6	*Alcaligenes faecalis, A. Eutrophus*	Organophosphate-chlorpyrifos	Hydrolysis and detoxification	[54]
7	*Pseudomonas* sp. BUR11	Parathion	Microbial degradation	[55]
8	*Flavobacterium sp.* ATCC27551 and *Escherichia coli* DH5α	Methyl parathion	Enzymatic hydrolysis and other metabolic pathways	[56]
9	*Rhodococcus sp. BCH2*	Atrazine	Biochemical degradation	[57]
10	*Bacillus* sp. SBER3	Polycyclic aromatic hydrocarbon (PAHs)	Rhizoremediation	[46]
11	*Arthrobacter fluorescens* and *Arthrobacter giacomelloi*	α, β, γ-hexachlorocyclohexane (HCH)	Bioremediation	[45]
12	*Arthrobacter nicotianae DH19*	Pentachloronitrobenzene (PCNB)	Bioremediation	[58]
13	*Sphingomonas.* strain F-7	Fevalerate	Hydrolysis of carboxyl ester linkage	[59]
14	*Burkholderia zhejiangensis*	Methyl parathion	Benzoquinone and benzenotrial pathway	[60]
15	*Alcaligenes faecalis* JBW4	Endosulfan	Non-oxidative pathways	[61]

(*Continued*)

Table 7.3 Different microbial species involved in organic pollutants degradation and their removal pathway. (*Continued*)

S. no.	Degrading microbes	Target organic pollutants	Removal pathways	References
16	*Comamonas testosteroni*	Naphthalene and Phenanthrene	Bioremediation and metabolic conversion	[62]
17	*Pseudomonas* sp. CF600 and *Stenotrophomonas maltophilia* KB2	4-Chlorophenol	Bioaugmentation	[43]
18	*Novosphingobium* strain DY4	2,4-Dichlorophenoxyacetic Acid	Bioaugmentation	[63]
19	*Rhodotorula* stain VITJzN03	Lindane	Dechloribationand oxidoruduction	[42]
20	*Acinetobacter sp. AVLB2*	4-nitroaniline	Rhizoremediation	[64]

pyrethroids, thereby improving plant productivity [48]. Beyond pesticides, bacteria also participate in the degradation of industrial dyes through enzymatic mechanisms and biosorption. *Pseudomonas aeruginosa* effectively enzymatically degrades azo-bonds, facilitating the breaking of persistent pollutants [49].

Recent studies have further highlighted the role of *Pseudomonas putida* strain Ch2, which effectively degrades caprolactam and glyphosate through glyphosate oxidoreductase pathways, contributing to, emphasizing its potential in the detoxification of xenobiotics [50] (Table 7.3). PGPR triggers plant growth through phytohormones such as auxins, cytokinins, and gibberellins, enhancing root development and nutrient uptake. Additionally, PGPR secretes enzymes such as dehydrogenases, peroxidases, and catalases that break down the organic pollutants [51]. Enzymatic degradation is further supported by oxidoreductases, a key group of enzymes that catalyze electron transport reactions essential for pollutant breakdown. These enzymes often require cofactors and involve the incorporation of molecular oxygen (O_2) [52]. To survive and function in pesticide-laden environments, microbes adapt by modifying receptor binding sites, mutating resistance genes, and overproducing detoxifying enzymes such as glutathione transferases and oxidases. For example, *Pseudomonas aeruginosa* degrades azo dyes *via* enzymatic reduction [49], while *Pseudomonas putida* aerobically metabolizes benzene, toluene, and xylene [44].

7.4 Future Remarks

To enhance the effectiveness of bioremediation, further research is crucial to unravel the mechanisms of biodegradation and biotransformation across diverse organisms such as plants, bacteria, fungi, and algae. In-depth studies are needed to characterize microbial enzymes involved in pesticide degradation and to uncover the unknown pathways facilitating HM uptake in plants. Equally important are investigations into root colonization efficiency, plant-microbe interaction, and the genetic basis of detoxification processes. Understanding site-specific microbial groups under varying environmental settings is essential, as challenges persist in selecting efficient microbial strains, managing environmental variables, and addressing the chemical complexity of contaminants. Progress in bioreactor design, the development of multistep biological treatment systems, and the formulation of effective microbial inoculants are also needed; additionally, gaining insights into plant mechanisms of contaminant accumulation, translocation, and detoxification – particularly in species that release chelating agents to recruit beneficial microbes – is vital. Many such plant species remain unexplored, especially in biodiversity-rich tropical regions. Finally, the influence of climate change on microbial dynamics and remediation efficiency calls for integrated studies to ensure the development of resilient and sustainable bioremediation strategies.

7.5 Recommendations

To advise sustainable remediation of HMs and POPs in agriculture, it is essential to deepen research on microbial and plant-based detoxification mechanisms, develop and field-validate efficient microbial consortia, and explore plant-microbe interactions that enhance stress resilience and pollutant degradation. Emphasis should be placed on site-specific microbial community characterization, climate-resilient bioremediation strategies, and scalable technology such as bioreactors and bioinoculants. Integrating PGPR-based approaches into eco-friendly agricultural practices, supported by strong policy frameworks, education, and stakeholder engagements, will be crucial for restoring soil health and confirming long-term farming sustainability.

7.6 Conclusion

Research on plant-microbe interactions is challenged by the complexity of rhizobacterial microbial communities, their dynamic interactions, and the influence of diverse biotic and abiotic stresses. Root exudates play a pivotal function in shaping microbial composition and activity within the rhizosphere. However, the extensive use of agrochemicals and high-yielding crop varieties has led to ecosystem degradation, compounded by the pressures of urbanization and industrialization on soil and water resources. Beneficial microbes, particularly PGPR, are essential for enhancing plant growth, minimizing agrochemical reliance, and facilitating the degradation of both organic and inorganic pollutants. The successes of sustainable agriculture rely on eco-friendly strategies, including microbial remediation of HMs and other contaminants. A deeper understanding of the molecular mechanisms governing these interactions is critical, as rhizospheric bacteria continue to show significant promise in restoring soil health and improving crop productivity.

List of Abbreviations

As	Arsenic
Cd	Cadmium
Cr	Chromium
CRISPR	Clustered regularly interspaced short palindromic repeats
Cu	Copper
DDT	Dichloro-diphenyl-trichloroethane
DEHP	Di(2-ethylhexyl) phthalate
DNA	Deoxyribonucleic acid
ECs	Emerging contaminants
HCH	Hexachlorocyclohexane
HMs	Heavy metals
MPs	Microplastics
N	Nitrogen
Ni	Nickel
$NiCl_2$	Nickel chloride
NT	Nanotechnology
O_2	Oxygen
PAHs	Polycyclic aromatic hydrocarbon
Pb	Lead

$Pb(NO_3)_2$	Lead nitrate
PCBs	Polychlorinated biphenyls
PCNB	Pentachloronitrobenzene
PCPs	Personal care products
PGPR	Plant Growth-Promoting Rhizobacteria
POPs	Persistent organic pollutants
ROS	Reactive oxygen species
SDGs	Sustainable Development Goals
UNSDGs	United Nations Sustainable Development Goals
V	Vanadium
Zn	Zinc

AI disclosure statement: The author confirms that no artificial intelligence (AI)-generated content or AI-assisted writing tools were used in the preparation of this chapter (including summaries and the abstract). All text, analysis, and conclusions are the result of the author's original work.

References

1. Paul, S., Singh, V., Chauhan, P.K., Srivastava, A.K., Upadhyay, S.K., Assessment of carrot growth performance with inoculation of AsT-PGPR under arsenic infested zone. *G-J. Environ. Sci. Technol.*, 7, 6, 78–84, 2020.
2. Kandpal, A., Singh, A., Jain, D., Upadhyay, S.K., Siddiqui, A.J., Sharma, K., Saxena, J., Managing plastic waste with nanotechnology: current sustainability prospects. *Nanotechnol. Environ. Eng.*, 8, 4, 1015–1031, 2023.
3. Alam, S., Bhardwaj, L.K., Mallick, R., Rai, S., Estimation of heavy metals and fluoride ion in vegetables grown nearby the stretch of River Yamuna, Delhi (NCR), India. *Indian J. Environ. Prot.*, 43, 1, 64–73, 2023.
4. Gogoi, B. and Sarma, H., Rhizoremediation of persistent organic pollutants (POPs) from the soil, in: *Land Remediation and Management: Bioengineering Strategies*, pp. 49–78, Springer Nature Singapore, Singapore, 2023.
5. Singh, B.N., Hidangmayum, A., Singh, A., Guru, A., Yashu, B.R., Singh, G.S., Effect of emerging contaminants on crops and mechanism of toxicity, in: *Sustainable Agriculture Reviews*, vol. 40, pp. 217–241, Springer International Publishing, Cham, 2020.
6. Upadhyay, S.K., Srivastava, A.K., Rajput, V.D., Chauhan, P.K., Bhojiya, A.A., Jain, D., Chaubey, G., Dwivedi, P., Sharma, B., Minkina, T., Root exudates: mechanistic insight of plant growth promoting rhizobacteria for sustainable crop production. *Front. Microbiol.*, 13, 916488, 2022.
7. Woldetsadik, D., Drechsel, P., Keraita, B., Itanna, F., Gebrekidan, H., Heavy metal accumulation and health risk assessment in wastewater-irrigated

urban vegetable farming sites of Addis Ababa, Ethiopia. *Int. J. Food Contam.*, 4, 1, 9, 2017.
8. Meharg, A.A. and Rahman, M.M., Arsenic contamination of Bangladesh paddy field soils: implications for rice contribution to arsenic consumption. *Environ. Sci. Technol.*, 37, 2, 229–234, 2003.
9. Kumar, S., Wang, M., Liu, Y., Zhu, Z., Fahad, S., Qayyum, A., Zhu, G., Vanadium stress alters sweet potato (*Ipomoea batatas* L.) growth, ROS accumulation, antioxidant defense system, stomatal traits, and vanadium uptake. *Antioxidants*, 11, 12, 2407, 2022.
10. Yang, J., Wang, M., Jia, Y., Gou, M., Zeyer, J., Toxicity of vanadium in soil on soybean at different growth stages. *Environ. Pollut.*, 231, 48–58, 2017.
11. Ashfaque, F., Inam, A., Iqbal, S., Sahay, S., Response of silicon on metal accumulation, photosynthetic inhibition and oxidative stress in chromium-induced mustard (*Brassica juncea* L.). *S. Afr. J. Bot.*, 111, 153–160, 2017.
12. Alharby, H.F. and Ali, S., Combined role of Fe nanoparticles (Fe NPs) and Staphylococcus aureus L. in the alleviation of chromium stress in rice plants. *Life*, 12, 3, 338, 2022.
13. Bagheri, R., Bashir, H., Ahmad, J., Iqbal, M., Qureshi, M.I., Spinach (*Spinacia oleracea* L.) modulates its proteome differentially in response to salinity, cadmium and their combination stress. *Plant Physiol. Biochem.*, 97, 235–245, 2015.
14. Jalil, S., Nazir, M.M., Al-Huqail, A.A., Ali, B., Al-Qthanin, R.N., Asad, M.A., Eweda, M.A., Zulfiqar, F., Onursal, N., Masood, H.A., Yong, J.W.H., Silicon nanoparticles alleviate cadmium toxicity in rice (*Oryza sativa* L.) by modulating the nutritional profile and triggering stress-responsive genetic mechanisms. *Ecotoxicol. Environ. Saf.*, 268, 115699, 2023.
15. Ahmed, T., Guo, J., Noman, M., Lv, L., Manzoor, N., Qi, X., Li, B., Metagenomic and biochemical analyses reveal the potential of silicon to alleviate arsenic toxicity in rice (*Oryza sativa* L.). *Environ. Pollut.*, 345, 123537, 2024.
16. Zeng, F., Ali, S., Zhang, H., Ouyang, Y., Qiu, B., Wu, F., Zhang, G., The influence of pH and organic matter content in paddy soil on heavy metal availability and their uptake by rice plants. *Environ. Pollut.*, 159, 1, 84–91, 2011.
17. Bhardwaj, L.K., Rath, P., Choudhury, M., Metzdorff, A., Jodder, P.K., Electronic wastes and contamination of water, in: *Electronic Waste*, pp. 48–57, CRC Press, Boca Raton, FL, USA, 2025.
18. Bhardwaj, L.K., Rath, P., Jain, H., Choudhury, M., Exploring the effects of e-waste on soil, water quality and human health. *Discov. Civ. Eng.*, 2, 1, 12, 2025.
19. de Freitas-Silva, L., de Araújo, T.O., da Silva, L.C., de Oliveira, J.A., de Araujo, J.M., Arsenic accumulation in Brassicaceae seedlings and its effects on growth and plant anatomy. *Ecotoxicol. Environ. Saf.*, 124, 1–9, 2016.
20. Vezza, M.E., Llanes, A., Travaglia, C., Agostini, E., Talano, M.A., Arsenic stress effects on root water absorption in soybean plants: Physiological and morphological aspects. *Plant Physiol. Biochem.*, 123, 8–17, 2018.

21. Al Mahmud, J., Hasanuzzaman, M., Nahar, K., Bhuyan, M.B., Fujita, M., Insights into citric acid-induced cadmium tolerance and phytoremediation in Brassica juncea L.: Coordinated functions of metal chelation, antioxidant defense and glyoxalase systems. *Ecotoxicol. Environ. Saf.*, 147, 990–1001, 2018.
22. Rafique, M.I., Ahmad, M., Al-Wabel, M.I., Ahmad, J., Al-Farraj, A.S., Mitigating the toxic effects of chromium on wheat (*Triticum aestivum* L.) seed germination and seedling growth by using biochar and polymer-modified biochar in contaminated soil. *Sustainability*, 14, 23, 16093, 2022.
23. Ma, J., Alshaya, H., Okla, M.K., Alwasel, Y.A., Chen, F., Adrees, M., Hussain, A., Hameed, S., Shahid, M.J., Application of cerium dioxide nanoparticles and chromium-resistant bacteria reduced chromium toxicity in sunflower plants. *Front. Plant Sci.*, 13, 876119, 2022.
24. Kiran, B.R. and Prasad, M.N.V., Biochar and rice husk ash assisted phytoremediation potentials of Ricinus communis L. for lead-spiked soils. *Ecotoxicol. Environ. Saf.*, 183, 109574, 2019.
25. Vasilachi-Mitoseru, I.C., Stoleru, V., Gavrilescu, M., Integrated assessment of Pb (II) and Cu (II) metal ion phytotoxicity on Medicago sativa L., Triticum aestivum L., and Zea mays L. plants: Insights into germination inhibition, seedling development, and ecosystem health. *Plants*, 12, 21, 3754, 2023.
26. Huang, J.S., Wei, D.P., Guo, X.Y., Ma, Y.B., Toxicity and accumulation of copper and nickel in wheat plants cropped on alkaline and acidic field soils. *Huan Jing Ke Xue= Huanjing Kexue*, 33, 4, 1369–1375, 2012.
27. Pathak, H.K., Chauhan, P.K., Seth, C.S., Dubey, G., Upadhyay, S.K., Mechanistic and future prospects in rhizospheric engineering for agricultural contaminants removal, soil health restoration, and management of climate change stress. *Sci. Total Environ.*, 927, 172116, 2024.
28. Singh, P., Chauhan, P.K., Upadhyay, S.K., Singh, R.K., Dwivedi, P., Wang, J., Jain, D., Jiang, M., Mechanistic insights and potential use of siderophores producing microbes in rhizosphere for mitigation of stress in plants grown in degraded land. *Front. Microbiol.*, 13, 898979, 2022.
29. Naz, H., Sayyed, R.Z., Khan, R.U., Naz, A., Wani, O.A., Maqsood, A., Maqsood, S., Fahad, A., Ashraf, S., Show, P.L., Mesorhizobium improves chickpea growth under chromium stress and alleviates chromium contamination of soil. *J. Environ. Manage.*, 338, 117779, 2023.
30. Akhtar, N., Ilyas, N., Yasmin, H., Sayyed, R.Z., Hasnain, Z., A., Elsayed, E., El Enshasy, H.A., Role of Bacillus cereus in improving the growth and phytoextractability of Brassica nigra (L.) K., Koch in chromium contaminated soil. *Molecules*, 26, 6, 1569, 2021.
31. Liu, A., Wang, W., Zheng, X., Chen, X., Fu, W., Wang, G., Ji, J., Jin, C., Guan, C., Improvement of the Cd and Zn phytoremediation efficiency of rice (Oryza sativa) through the inoculation of a metal-resistant PGPR strain. *Chemosphere*, 302, 134900, 2022.

32. Bruno, L.B., Anbuganesan, V., Karthik, C., Kumar, A., Banu, J.R., Freitas, H., Rajkumar, M., Enhanced phytoextraction of multi-metal contaminated soils under increased atmospheric temperature by bioaugmentation with plant growth promoting Bacillus cereus. *J. Environ. Manage.*, 289, 112553, 2021.
33. Dhaliwal, S.S., Sharma, V., Shukla, A.K., Verma, V., Kaur, M., Shivay, Y.S., Nisar, S., Gaber, A., Brestic, M., Barek, V., Skalicky, M., Biofortification—A frontier novel approach to enrich micronutrients in field crops to encounter the nutritional security. *Molecules*, 27, 4, 1340, 2022.
34. Chauhan, P.K., Upadhyay, S.K., Rajput, V.D., Dwivedi, P., Minkina, T., Wong, M.H., Fostering plant growth performance under drought stress using rhizospheric microbes, their gene editing, and biochar. *Environ. Geochem. Health*, 46, 2, 41, 2024.
35. Sharma, S. and Saraf, M., Biofilm-forming plant growth-promoting rhizobacterial consortia isolated from mines and dumpsites assist green remediation of toxic metal (Ni and Pb) using Brassica juncea. *Biol. Futura*, 74, 3, 309–325, 2023.
36. Belhassan, M., Farhat, A., Abed, H.E., Chaabeen, Z., Bouzid, F., Elleuch, A., Fendri, I., Khemakhem, B., Isolation and identification of a new Bacillus glycinifermentans strain from date palm rhizosphere and its effect on barley seeds under heavy metal stress. *Braz. J. Microbiol.*, 55, 1, 843–854, 2024.
37. Ali, J., Ali, F., Ahmad, I., Rafique, M., Munis, M.F.H., Hassan, S.W., Sultan, T., Iftikhar, M., Chaudhary, H.J., Mechanistic elucidation of germination potential and growth of Sesbania sesban seedlings with Bacillus anthracis PM21 under heavy metals stress: An *in vitro* study. *Ecotoxicol. Environ. Saf.*, 208, 111769, 2021.
38. Bruno, L.B., Karthik, C., Ma, Y., Kadirvelu, K., Freitas, H., Rajkumar, M., Amelioration of chromium and heat stresses in Sorghum bicolor by Cr6+ reducing-thermotolerant plant growth promoting bacteria. *Chemosphere*, 244, 125521, 2020.
39. Liu, A., Wang, W., Chen, X., Zheng, X., Fu, W., Wang, G., Ji, J., Guan, C., Phytoremediation of DEHP and heavy metals co-contaminated soil by rice assisted with a PGPR consortium: Insights into the regulation of ion homeostasis, improvement of photosynthesis and enrichment of beneficial bacteria in rhizosphere soil. *Environ. Pollut.*, 314, 120303, 2022.
40. Syed, A., Elgorban, A.M., Bahkali, A.H., Eswaramoorthy, R., Iqbal, R.K., Danish, S., Metal-tolerant and siderophore producing *Pseudomonas fluorescens* and Trichoderma spp. improved the growth, biochemical features and yield attributes of chickpea by lowering Cd uptake. *Sci. Rep.*, 13, 1, 4471, 2023.
41. Altaf, M., Ilyas, T., Shahid, M., Shafi, Z., Tyagi, A., Ali, S., Trichoderma inoculation alleviates Cd and Pb-induced toxicity and improves growth and physiology of *Vigna radiata* (L.). *ACS Omega*, 9, 7, 8557–8573, 2024.

42. Abdul Salam, J., Lakshmi, V., Das, D., Das, N., Biodegradation of lindane using a novel yeast strain, Rhodotorula sp. VITJzN03 isolated from agricultural soil. *World J. Microbiol. Biotechnol.*, 29, 3, 475–487, 2013.
43. Nowak, A. and Mrozik, A., Degradation of 4-chlorophenol and microbial diversity in soil inoculated with single Pseudomonas sp. CF600 and Stenotrophomonas maltophilia KB2. *J. Environ. Manage.*, 215, 216–229, 2018.
44. Mukherjee, A.K. and Bordoloi, N.K., Biodegradation of benzene, toluene, and xylene (BTX) in liquid culture and in soil by Bacillus subtilis and Pseudomonas aeruginosa strains and a formulated bacterial consortium. *Environ. Sci. Pollut. Res.*, 19, 8, 3380–3388, 2012.
45. De Paolis, M.R., Lippi, D., Guerriero, E., Polcaro, C.M., Donati, E., Biodegradation of α-, β-, and γ-hexachlorocyclohexane by Arthrobacter fluorescens and Arthrobacter giacomelloi. *Appl. Biochem. Biotechnol.*, 170, 3, 514–524, 2013.
46. Bisht, S., Pandey, P., Kaur, G., Aggarwal, H., Sood, A., Sharma, S., Kumar, V., Bisht, N.S., Utilization of endophytic strain Bacillus sp. SBER3 for biodegradation of polyaromatic hydrocarbons (PAH) in soil model system. *Eur. J. Soil Biol.*, 60, 67–76, 2014.
47. Philip, L., Bioremediation of endosulfan contaminated soil and water—Optimization of operating conditions in laboratory scale reactors. *J. Hazard. Mater.*, 136, 2, 354–364, 2006.
48. Chen, S., Chang, C., Deng, Y., An, S., Dong, Y.H., Zhou, J., Hu, M., Zhong, G., Zhang, L.H., Fenpropathrin biodegradation pathway in Bacillus sp. DG-02 and its potential for bioremediation of pyrethroid-contaminated soils. *J. Agric. Food Chem.*, 62, 10, 2147–2157, 2014.
49. Sarayu, K. and Sandhya, S., Aerobic biodegradation pathway for Remazol Orange by Pseudomonas aeruginosa. *Appl. Biochem. Biotechnol.*, 160, 4, 1241–1253, 2010.
50. Esikova, T.Z., Anokhina, T.O., Suzina, N.E., Shushkova, T.V., Wu, Y., Solyanikova, I.P., Characterization of a new Pseudomonas Putida strain Ch2, a degrader of toxic anthropogenic compounds epsilon-Caprolactam and glyphosate. *Microorganisms*, 11, 3, 650, 2023.
51. Takio, N., Yadav, M., Yadav, H.S., Catalase-mediated remediation of environmental pollutants and potential application–a review. *Biocatal. Biotransform.*, 39, 6, 389–407, 2021.
52. Bhandari, S., Poudel, D.K., Marahatha, R., Dawadi, S., Khadayat, K., Phuyal, S., Shrestha, S., Gaire, S., Basnet, K., Khadka, U., Parajuli, N., Microbial enzymes used in bioremediation. *J. Chem.*, 2021, 1, 8849512, 2021.
53. Chen, Z., Jiang, Y., Chang, Z., Wang, J., Song, X., Huang, Z., Chen, S., Li, J., Denitrification characteristics and pathways of a facultative anaerobic denitrifying strain, Pseudomonas denitrificans G1. *J. Biosci. Bioeng.*, 129, 6, 715–722, 2020.

54. Yadav, U., Kushwaha, S., Anand, V., Kumar, S., Prakash, O., Singh, P.C., Chlorpyrifos degradation by plant growth-promoting *Alcaligenes faecalis* bacteria isolated from oil-contaminated soil. *Biorem. J.*, 25, 1, 80–88, 2020.
55. Pailan, S. and Saha, P., Chemotaxis and degradation of organophosphate compound by a novel moderately thermo-halo tolerant Pseudomonas sp. strain BUR11: evidence for possible existence of two pathways for degradation. *PeerJ*, 3, e1378, 2015.
56. Abdel-Razek, M.A.R.S., Folch-Mallol, J.L., Perezgasga-Ciscomani, L., Sánchez-Salinas, E., Castrejón-Godínez, M.L., Ortiz-Hernández, M.L., Optimization of methyl parathion biodegradation and detoxification by cells in suspension or immobilized on tezontle expressing the opd gene. *J. Environ. Sci. Health, Part B*, 48, 6, 449–461, 2013.
57. Kolekar, P.D., Phugare, S.S., Jadhav, J.P., Biodegradation of atrazine by *Rhodococcus sp. BCH2* to N-isopropylammelide with subsequent assessment of toxicity of biodegraded metabolites. *Environ. Sci. Pollut. Res.*, 21, 3, 2334–2345, 2014.
58. Wang, Y., Wang, C., Li, A., Gao, J., Biodegradation of pentachloronitrobenzene by Arthrobacter nicotianae DH19. *Lett. Appl. Microbiol.*, 61, 4, 403–410, 2015.
59. Yu, F.B., Shan, S.D., Luo, L.P., Guan, L.B., Qin, H., Isolation and characterization of a Sphingomonas sp. strain F-7 degrading fenvalerate and its use in bioremediation of contaminated soil. *J. Environ. Sci. Health, Part B*, 48, 3, 198–207, 2013.
60. Popoca-Ursino, E.C., Martínez-Ocampo, F., Dantán-González, E., Sánchez-Salinas, E., Ortiz-Hernández, M.L., Characterization of methyl parathion degradation by a Burkholderia zhejiangensis strain, CEIB S4–3, isolated from agricultural soils. *Biodegradation*, 28, 5, 351–367, 2017.
61. Kong, L., Zhu, S., Zhu, L., Xie, H., Su, K., Yan, T., Wang, J., Wang, J., Wang, F., Sun, F., Biodegradation of organochlorine pesticide endosulfan by bacterial strain *Alcaligenes faecalis* JBW4. *J. Environ. Sci.*, 25, 11, 2257–2264, 2013.
62. Lu, Q., Sun, X., Jiang, Z., Cui, Y., Li, X., Cui, J., Effects of Comamonas testosteroni on dissipation of polycyclic aromatic hydrocarbons and the response of endogenous bacteria for soil bioremediation. *Environ. Sci. Pollut. Res.*, 29, 54, 82351–82364, 2022.
63. Dai, Y., Li, N., Zhao, Q., Xie, S., Bioremediation using Novosphingobium strain DY4 for 2, 4-dichlorophenoxyacetic acid-contaminated soil and impact on microbial community structure. *Biodegradation*, 26, 2, 161–170, 2015.
64. Silambarasan, S. and Vangnai, A.S., Biodegradation of 4-nitroaniline by plant-growth promoting Acinetobacter sp. AVLB2 and toxicological analysis of its biodegradation metabolites. *J. Hazard. Mater.*, 302, 426–436, 2016.

8

Socioeconomic and Policy Considerations

Samarendra Singh[1], Saloni Guleria[1], Aditya Raj[1], Gautam Singh[1] and Navjot Kaur Sandhu[2*]

[1]*Department of Pharmacy Practice, ISF College of Pharmacy, Moga, Punjab, India*
[2]*Department of Quality Assurance, ISF College of Pharmacy, Moga, Punjab, India*

Abstract

Heavy metals (HMs) contamination is not simply an environmental or a technological problem, but a firmly rooted socioeconomic and policy issue. Sometimes the remediation initiatives are successful or fail based on how much they are integrated with the social, economic, and governance facets, combined with scientific and technical interventions. The current study criticizes the socioeconomic and policy processes that influence the process of planning, implementing, and achieving a long-term sustainability of HM remediation initiatives. The effects of HM pollution were, in many cases, disproportionately borne by vulnerable portions of the population, as well as those already economically disadvantaged, leading to inequity in health and livelihood impacts. This study examines the interaction between economic determinants, which include remediation costs, funding availability, and land-use conflicts, and how it affects the livelihood of communities, and the political and institutional structures that shape environmental governance. It explores regulatory frameworks, enforcement, and economic instruments such as the polluter-pays principle (PPP), tax credits, and participation by the general public, for example, and how those either facilitate or prevent the cleanup of the environment. The key themes are community involvement, openness, and participatory decision-making as a prerequisite to trust and sustainable results. Based on case studies in diverse geopolitical conditions, this study outlines how socioeconomics and policy environments shape remediation strategies, promotes best practices that combine environmental, social, and economic objectives, and argues that a multidisciplinary, policy-side approach based on local circumstances and robust governance is the way forward.

**Corresponding author*: navjotsandhu9619@gmail.com; ORCID: https://orcid.org/0000-0002-6295-1798

Prangya Rath, Laxmi Kant Bhardwaj and Amit Kumar (eds.) Heavy Metal Remediation, (195–214)

***Keywords*:** Environmental justice, governance, heavy metals, public participation, remediation policy, socioeconomic impact

8.1 Introduction

Heavy metal (HM) contamination is one of the most severe environmental problems that has ever plagued the 21st century, with serious and far-reaching impacts on the health and well-being of people, the integrity and stability of the natural environments, and the success and prosperity of human development. The HMs, such as lead (Pb), cadmium (Cd), mercury (Hg), and arsenic (As), are naturally occurring elements, but have become popular despite human efforts, as sources of environmental pollution. They are particularly hazardous to both human and environmental health since they are non-biodegradable and bioaccumulative in nature [1, 2]. Unlike most organic pollutants, HMs are not degradable into other less dangerous substances, resulting in adverse long-term and sometimes reversible effects on soil, water bodies, flora, and fauna, and eventually human beings.

The increasing awareness regarding the socioeconomic aspects of HM pollution indicates an important departure from mere scientific and technical discourses to those that address social inequalities and challenges, regulation, and social interactions. The most overstrained by the pollution of HMs are the people who are socioeconomically disadvantaged, i.e., low-income groups, indigenous population, women, children, and people working in the pollutant industries, such as the mining and manufacturing businesses. These groups tend to be over-represented amongst the targets of pollution due to their proximity to the polluting activity, and in most instances, they do not have adequate access to health care and do not have adequate legal or political points of contact.

Nothing to do with picking up HM pollution, then, can be regarded as restricted to the field of environmental science or engineering. It requires holistic solutions that make use of the socioeconomical real environment, the essence of environmental justice, and effective policy and governing systems. The process of sustainable remediation needs to be approached on a participatory basis involving appropriate affected communities, and transparent, accountable decision-making needs to be ensured, and equity needs to be built into the policy tools and the institutional settings [3].

The given chapter inspects essential socioeconomic and policy issues in the context of HM remediation. These goals are several-fold: to uncover the profound and multifaceted socioeconomic consequences on health, livelihood, and social equity of HM contamination; to discuss the principles of environmental justice in terms of disproportionate burden and empowerment to remediation; to examine current international and national policy frameworks and governance institutions defining remediation; and, lastly, to propose how the public can participate and how structures of socially equitable programs at reasonable costs can be designed. In conceptualizing HM remediation in this holistic, multidisciplinary framework, this chapter will seek to make a contribution to a more humane, equitable, and sustainable research agenda that connects science, technology, and social governance to the healing of both the polluted ocean and land as well as people.

8.2 Socioeconomic Impacts of Heavy Metal (HM) Contamination

8.2.1 Health Impacts and Vulnerable Populations

Contaminated food, water, and air are the main ways HMs enter human bodies, but they can also do so directly through work. Toxicological effects vary according to metal type, the dose, duration of exposure, and individual sensitivities [4]. Vulnerable populations, including children, pregnant women, the elderly, populations with indigenous backgrounds, and populations living in low-income societies, are over-disproportionately affected by the combination of physiological and social vulnerability [3]. Exposure to Pb, usually due to contaminated drinking water, soil, and deteriorating Pb-based paints, poses irreversible harm in the developing brain and nervous system of children, with the consequences of lowering intelligence quotient (IQ), behavioral disorders, and learning disabilities [5].

Chronic Pb exposure increases cardiovascular risks and kidney damage in the case of adults [6]. Cd, which is widely present in batteries and industrial effluents, is highly toxic to renal functions and bone metabolism. With chronic exposure, there is an association with osteoporosis, kidney disease, and the risk of cancer [7]. Hg is enriched and released on food webs, such as in regions of methylmercury, and the consumption of fish is an important exposure pathway. The incidence of Hg poisoning impacts the neurological aspects, where fetal development is at an increased risk of cognitive deficiencies and motor problems [8]. Exposure to As, largely by way of groundwater contamination in nations like Bangladesh, India, and some regions in China, results in skin lesions, heart diseases, and cancer of the skin, lung, and bladder [9, 10]. The poisonous impacts are layered by the presence of co-exposures and compounded stressors experienced by vulnerable populations and include poverty, low access to health services,

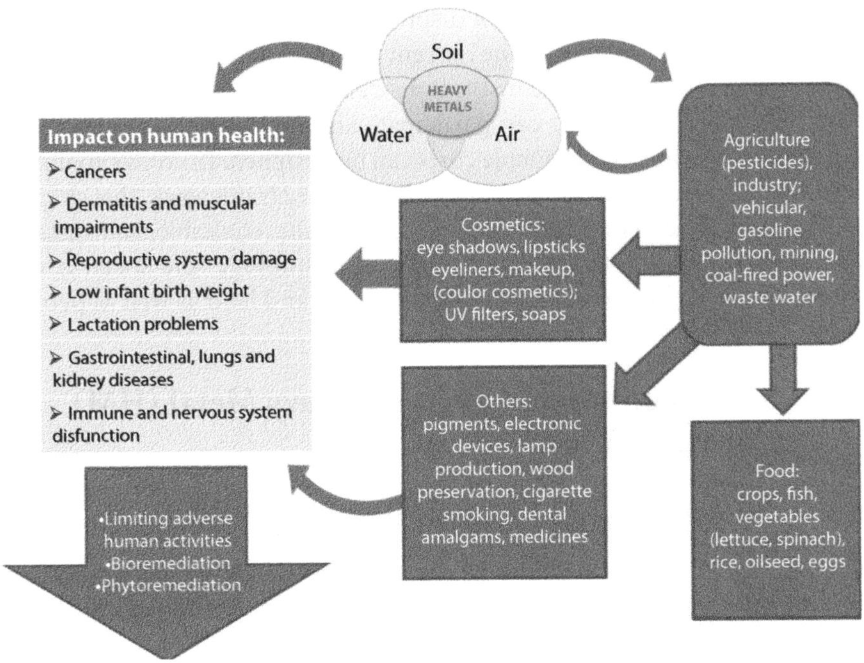

Figure 8.1 Routes of exposure, the impact of toxic metals on human health, and the ways of limiting the risk caused by contact with these elements (large arrow on the left). These adverse effects are caused by direct exposure to toxic metals in the environment or indirectly due to anthropogenic activity.

and poor nutrition. Unequal distribution of roll of risks and undesirable health outcomes is heartbreaking as well as a good example of environmental injustice, where social injustice and toxicological reality collide. Figure 8.1 shows the routes of exposure, impacts of metals on the health of humans, and ways of limiting the risk.

8.2.2 Economic Costs and Livelihood Disruption

The economic costs of HM contamination, by far, surpass health care in all other aspects of livelihoods, productivity, and community resilience. Pollution of agricultural soils impairs agricultural production and may make agricultural products unsafe to use and trade, hence threatening food security and income-earning groups, particularly smallholder farmers who solely rely on farming [11]. Polluted water pipes reduce fisheries and livestock production, further increasing the imbalance in the rural economies [12]. Treatment relating to HMs to the disease and its related chronic diseases and disability is a direct medical expenditure that brings financial implications in households and health budgetary requirements of the nation [13]. Economic growth is also impeded by indirect costs that occur; these include lost labor productivity and sick absenteeism. Moreover, devaluation and stigmatization of the properties in contaminated sites are frequent, alienating investment attractors and tax bases to the municipalities. Contamination may also affect regional development, as social tensions and migration can increase due to the disruptions of the social and economic systems [14].

8.2.3 Social Inequities and Environmental Justice

Socio-spatial patterns of HM pollution are commonly replicable and supportive of the social orders and discrimination. Unwanted locations lie close to marginalized or politically disenfranchised communities, because they do not usually have the resources or power to fight pollution or to prevent pollution by encouraging demands in order to protect their health. The patterns have been motivated by poor legal protection, below par transparency, and long-term disregard by designating them as sacrifice zones-where environmental harm and social lack converge. Part of the resulting disparities may be perpetuated unless environmental governance is explicitly concerned with equity issues, as elsewhere in remediation activity. Environmental justice frameworks strive to right this by giving primacy of voice to disadvantaged communities, equalizing the giving and taking of environmental costs and benefits, and institutionalizing procedural fairness by providing opportunities to participate in decision-making.

8.3 Environmental Justice in Heavy Metal (HM) Remediation

8.3.1 Disproportionate Impacts on Marginalized Groups

The question of environmental justice is a pivotal perspective within which we can view and explore how HM contamination has adversely impacted vulnerable populations in most parts of the world. In varied contexts, social, economic, and political inequalities intersect to cause elevated exposure and vulnerability amidst communities that are ordinarily marginalized in decisions. The indigenous people and rural people in most developing countries carry the excess weight of being adjacent to the mining area and industrialized centers [15]. Such groups often have no formal land rights and thus are more exposed to eviction/land alienation and environmental changes. As one example, artisanal and small-scale mining is a significant livelihood activity in some Sub-Saharan African countries, where it is characterized by lax safety and environmental standards. The communities adjacent to the contaminated river systems tend to live under potentially hazardous exposure to Hg and Pb levels in their bodies; hence, the instant rise in infant mortality as well as the chronic diseases.

Existing or abandoned industrial facilities are regularly hotspots of legacy pollution, usually located in low-income and minority populations within urban America. The issue of systemic negligence can be considered a harsh lesson in the aftermath of the United States (US) Flint water crisis, when mainly African American communities were vulnerable to Pb-poisoned water sources after financial belt-tightening procedures and governance failure. These are some examples of how institutional racism and economic marginalization produce a direct translation of this into environmental health crises. Discriminated groups routinely encounter obstacles to healthcare access, legal recourse, and risk information as well. This exposure causes exacerbation of the health effects of the HMs and communication of the cleanup process. Further, social reaction to pollution can also look complex, with economic dependence on polluting industries that lead to communities measuring short-term job opportunities against the effects of the damaged environment [16, 17].

8.3.2 Case Studies Illustrating Environmental Injustice

- ***Sub-Saharan Africa: Communities in Artisanal Mining***

Artisanal gold (Au) mining in nations such as Ghana, Zambia, and Burkina Faso employs millions of people and tends to use Hg in the process of extracting Au, and there is minimal regulation. The third world has local populations exposed to water sources and soil pollution, which are coupled with poverty and a paucity of healthcare systems. Exposure to Hg causes neurological impairments, particularly to children, and the results are disastrous socially [18].

- ***The United States (US) Urban Lead (Pb) Poisoning***

Several urban communities in the US, such as Baltimore, Chicago, and Detroit, have recorded high blood levels of Pb, mostly in the low-income minority communities. Residential segregation, old infrastructure with

Pb-based colorings, and industrial zoning regulations are elements that behave as a source of ongoing nonsymmetry. An upsurge in the health response to the environment has ensued, even though there are gaps; still, there is an advocated need for more equitable environmental governance.

- ***South and East Asian Industrial Pollution***

Blistering industrialization of India and China has also caused the localization of HM release and could severely affect millions of agricultural families. In regions such as Punjab (India) and Pearl River Delta (China), pre-contaminated groundwater of As and Cd poses a risk to contamination of food and diseases. These issues are compounded by the failure to enforce regulatory action and the lack of involvement of the people [16, 17].

8.3.3 Principles of Environmental Justice in Remediation

In circumstances of environmental justice-oriented remediation activities, equity and democratic involvement are prioritized by giving priority to historically burdened communities, empowering the local stakeholders, establishing transparency, and enforcement of health surveillance and compensation. Remediation methods come with a clear identification and focus on the communities that have been disproportionately harmed, making information, technical support, and chances to affect the decision-making process available. Transparency in the communication of risks, remediation plans, and progress leads to trust and accountability. Mechanisms that monitor health effects and redress to injured parties are useful to combat continuing harms and the creation of social legitimacy.

8.4 Policy and Regulatory Frameworks

8.4.1 International and National Legal Frameworks

The problem of HM contamination is a global one that triggered the creation of several international frameworks that can guide and coordinate the reduction of pollution and the introduction of rehabilitation. The most prominent of these is the Minamata Convention on Hg, a multilateral treaty created in 2013 to govern Hg emissions and Hg releases, with guided implementation as seen in guidelines of Hg supplies and the use of Hg in products and processes, and disposal of Hg wastes [19]. The convention states expressly that there is a need to protect the vulnerable population and ecosystems against exposure to Hg. To supplement this, the Basel Convention (Basel Convention on the Control of Transboundary Movements of Hazardous Wastes and Their Disposal) controls the international traffic of hazardous wastes such as HMs [20]. It helps nations to deal with polluted fills properly to avert environmental dumping in susceptible areas.

Nationally, a country formulates laws and regulations that determine the maximum tolerance of the concentrations of HMs in air, soil, water, and food. They put liability and liability frameworks in place to make contamination and remediation liability, as well as to set standards of clean-up technologies. An example is the Comprehensive Environmental Response, Compensation, and Liability Act (CERCLA), popularly known as Superfund, because it allows contaminated sites to be located and cleaned up, and legal action to be taken against the responsible parties in the US. There are differences, however, in the breadth and effectiveness with which the national regulations have been adopted. Advanced economies tend to have higher standards and supervision capabilities, and developing economies tend to be confronted by infrastructural, institutional, and financing realities that strangle regulation [21].

8.4.2 Effectiveness and Limitations of Current Policies

Though the international treaties and the national regulations can serve as the foundation of actions, there are a number of difficulties that can be seen as the limiting factors of their use, such as poor enforcement, jurisdictional fragmentation, low rate of public participation, and transboundary pollution. Insufficient regulatory capacity is rampant in countries, resulting in ineffective monitoring, disparate inspection visits, and poor enforcement with respect to violations. There can be a problem of overlapping mandates by multiple agencies, breeding confusion and inefficiencies, as well as leaving gaps in how to deal with contamination more comprehensively. Policy formulation and its translation do not always have adequate and effective participation of the affected community, thereby lowering its acceptance and openness. Emissions of HMs transfer between countries across air and water, requiring the development of regional approaches that are poorly developed. The above challenges are emerging as a major concern to strengthen institutional frameworks, align regulations, and increase community participation so that policies are achieved through concrete results in the form of remediation activities [22].

8.5 Governance and Institutional Arrangements

8.5.1 Stakeholder Roles

Remediation of HMs is the area that calls on the concerted effort and duties of various stakeholders. They contrast and yet depend on each other to guarantee sustainable and fair cleanup. The national, regional, and local governments are mandated with duties of identifying regulatory guidelines, implementing environmental standards, funding remediation activities, and running health care. The agencies involved in the environment, health, industry, and community development should coordinate to respond in a coordinated fashion. Businesses that manufacture, dispose of waste, or otherwise generate the contaminant forms of HMs have the obligation of keeping pollution, switching to cleaner technologies, and funding or otherwise managing remedial efforts on the basis of the PPP. Newer technologies to remediate can also come into play as innovations in the private sector. Non-Governmental Organizations (NGOs) are regarded as having a lot of potential to participate in advocacy, monitoring, and awareness creation of pollution mobilization of the people in the affected areas, and accountability by governments and businesses. They are inclined to provide connections between the technical specialists and the local stakeholders [23]. Local and indigenous communities have important local knowledge and traditional knowledge about changes and effects on the environment and health because they are situated near the points of pollution. They play a major role in establishing priorities of remediation, solution-provision, and follow-up of results [23].

8.5.2 Transparency, Accountability, and Enforcement

The governance performance depends on open processes and accountable agencies. Transparency is the possibility of the population obtaining pollution information, legislation, information on regulations, and the decision-making process. Openness aids in developing trust among all stakeholders, especially those who were marginalized or left out in history. Laws that impose liability on the polluter, third-party reviews on the nature of remediation, and effective systems of grievance or appeal are the accountability mechanisms. There must be enough financial, technical, and human resources to be utilized in the inspection, follow-up, and penalization of the violations. These, otherwise, appear to lead to ineffective or decorative policies [23].

8.5.3 Collaborative Governance Models

Remediation governance is becoming more participatory in nature, with a range of actors participating in joint decision-making and action. Multi-stakeholder task forces mean that government, industry, scientists, NGOs, and community leaders are included in the task force, thus there is assured diversity of input in policy and practice. Giving individuals in a location the capability and the skills necessary to monitor ecological conditions will present advanced notification of pollution, in addition to creating awareness. Partnerships that draw upon efficiencies of the private sector and its innovation, and retain governmental accountability, build sustainability, and allow a larger remediation scope. The governance of remediation is becoming more communal, where the diverse stakeholders are engaged in common decision-making and taking [23].

8.6 Public Participation in Remediation Efforts

8.6.1 Importance of Community Involvement

The success and sustainability of HM remediation activities depend on their efforts to obtain the input of the populace. The involvement in remediation activities makes affected communities increase the legitimacy, relevance, and effectiveness of actions taken. The communities also offer vital domestic knowledge on the sources of contamination, exposure pathways, and social networks, which can be ignored by outsiders. Their participation is conducive to assuring that priorities relate to local requirements and values in remediation. In addition, community engagement helps increase the degree of transparency and faithfulness between interested persons, minimizing altercations and opposition risks that may jeopardize the projects or delay them. Participating of residents in the problem/risk assessment, choice of technology to be used, and their monitoring, will encourage a sense of ownership of solutions and long-term stewardship [23].

8.6.2 Methods and Challenges

Effective HM remediation is anchored in meaningful participation of the community by making them active participants in the ecological management process and therefore beneficiaries. There is a variety of participation methods, and all of them are strong in different situations, but they all have the same goal of encouraging dialogue, transparency, and shared responsibility. In addition to these approaches, there should be other issues that need to be identified and addressed to come up with a true realization of community participation.

8.6.3 Methods Facilitating Public Participation

8.6.3.1 Public Hearings and Consultations

Formalized gatherings such as public hearings and consultations are organized so that the authorities can present important information on the dangers of contamination, the mitigation strategy, the time frame, and the projected results. These discussion groups encourage individuals within the community to give feedback, voice concerns, and put forth questions. These are usually required as part of any regulatory framework and can be used at various points in the project's lifecycle to incorporate the voice of the people into the planning or implementation processes. The strategy acts as a critical communication linkage among professionals, policymakers, and locals. It also gives honest information to communities, written down, the rumors and misinformation are removed, and then a base is provided where people can ventilate their grievances. To ensure the effective functioning of hearings, the locations will have to be comfortable, publicize the hearing beforehand, and work with qualified facilitators to encourage people to attend the hearings and manage contrasting opinions [23].

8.6.3.2 Community Advisory Boards (CABs)

Community advisory boards (CABs) are formal groups of community leaders, local health professionals, activists, or even representatives of the government or company, with regular meetings intended to address any issues of the remediation process and act as a watchdog. These are standing liaison boards that are used to advise the remediation process, to indicate to the agencies the priorities of the community, as well as dispelling misunderstandings or narrowing down conflicts. CABs make sure that the development of trust is achieved through institutionalizing channels of participation beyond the rare hearing. They facilitate long-term communication and flexibility, and the communities achieve flexibility and adapt in response to emergent issues or remediation strategy alterations. Effective CABs are more inclusive, balanced in representation, and are found to share their information transparently.

8.6.3.3 Participatory Mapping and Risk Assessment

Participatory mapping means education and engagement of residents in marking contamination hotspots, pathways of exposure, and sensitive areas with the use of visual aids, such as maps, diagrams, or online solutions. The complementary participatory risk assessment approach involves community members in activities aimed at determining health symptoms as well as the estimation of the exposure frequency, blending the scientific scales with the lived experiences. These collaborative technologies tap into the area-specific knowledge that outsider specialists can otherwise never gain access to, so as to unmask dormant contamination sectors or areas of social susceptibility. They boost legitimacy by making the community insights legitimate ones and the sense of ownership of data collection, as well as interpretation. There is also the aspect of participatory processes, democratizing scientific processes by overcoming technical-expert rifts.

8.6.3.4 Citizen Science and Monitoring

Using community members as their own environmental data collectors in the form of citizen scientists, making the community members gather, say, a soil sample, or air quality readings, or test water, offers a compelling engagement and monitoring tool. Through training programs, residents should be informed of the basic scientific abilities to guide them to observe the amount of pollution, new sources of contaminants, and check the remediation competency. Citizen science creates a sense of empowerment and helps in converting mere observers into active participants in environmental good practices. The surveillance can be complemented with data that is gathered, which leads to quicker action and improves accountability. Effective programs guarantee technical assistance, quality assurance, and transparency regarding the conduct of conclusions into a policy [23].

8.6.3.5 Grievance Mechanisms

Proper grievance/complaint mechanisms should be accessible to people showing concern over remediation processes, health effects, or social tensions. Good systems have as many points of entry as possible-hotlines, suggestion boxes, neighborhood meetings- and respond quickly by investigating complaints and making decisions. The grievance system is evidence of institutional responsibility and dedication to the neighborhood's good. They manage to decrease tensions, prevent the deterioration of conflict, and enhance trust. It includes significant design decisions, or aspects, which are supposed to have a long list of beneficial effects, i.e., a higher rate of usage and its effectiveness: anonymity, cultural sensitivity, and procedural clarity.

8.6.4 Challenges to Meaningful Participation

Regardless of such diverse and complementary ways of engaging communities, the major barriers are likely to hinder a considerably successful and viable community participation.

8.6.4.1 Technical Complexity

The scientific and regulatory terminologies associated with HM contamination and remediation are very technical, and they are entrenched with jargon, risk assessments, and risk legislation. This technical complexity has the potential to repel non-experts, rendering communities lost or overpowered. Conscious translation of technical information to simple, culturally acceptable, and non-jargon language is essential to exclude and disempower the residents. These further strains misunderstanding and lowers the quality of participation.

8.6.4.2 Information Access

The main characteristic of successful participation is fair access to appropriate information. Understanding among many of the affected communities, however, may be hampered by literacy levels and differences in language, and a scarcity of domestic communication infrastructure. Further impairment of access is based on poor distribution methods, restricted outreach methods, ill-timed announcements, or the absence of multi-media strategies. Addressing the gaps means serious investment in the education process, culturally appropriate content, and communication approaches involving local languages, visual representations, and oral traditions.

8.6.4.3 Mistrust

Past patterns of neglect, or unfulfilled promises or exploitative relationships between authorities, companies, and communities, produce profound doubts. Such suspicion can stop community members from getting involved in the process of participation, which may be viewed as superficial or tokenistic. The process of building trust within a community is a lengthy process based on transparency, responsiveness, and a track record of listening to voices within the community. Not confronting mistrust endangers the risk of sidelining the same groups that participation efforts intend to empower.

8.6.4.4 *Social Exclusion*

The presence of the activities of the marginalized groups, quite often, manifests itself through problems in attending them, starting with women, ethnic minorities, indigenous people, the elderly, and economically disadvantaged individuals. Their exclusion may be because of social norms, gender, or power structure, or even culture against some of the formalized procedures, i.e., public hearings or advisory boards. Inclusion is key and requires activating outreach, provision of a safe internal space, and specific capacity building. Unless these structural exclusions are addressed, the process of participation may represent nothing more than interest and interests and serve as a vehicle to recreate systemic inequalities [23].

8.7 Economic Valuation Methods

The total picture of socioeconomic aspects of HM remediation requires quantifying the direct costs and benefits, not only the direct financial costs and benefits, but also social and environmental values. Seasons adjust to this as methods of economic valuation adjust to this, and they are cost-benefit analysis (CBA), social return on investment, contingent valuation, willingness to pay, and ecosystem service valuation. CBA is a method that weighs total costs of remediation initiatives, which are usually presented as costs of acquiring technology, labor, and maintenance against the posited benefits of, for example, better health, productivity, and benefits to ecosystem services. Social return on investment goes beyond the narrow dollar costs to ensure social, environmental, and economic benefits, and documents the broader societal benefits that result from the cleanups. Contingent valuation and willingness-to-pay are survey-based methods to determine the monetary value that communities place on environmental quality (or the value they place on reduced toxic risks to human health). The estimated savings of using restored ecosystems to filter water, soil nutrition, and support biodiversity are all essential measures in areas that depend on natural resources. The combination of these methods explains the decision-makers based on a comprehensive/holistic view of remediation premiums without downplaying investment due to limited cost consideration.

The remediation of HMs may have multidimensional outcomes such as health gains, economic development, social cohesion, and environmental sustainability. The decrease in the rates of chronic diseases, disability, and death means a decline in the individual and societal health spending, which enhances health. Rehabilitation of farmlands, fisheries, and business places helps revive local economies and the number of local jobs. Open and participatory remedial works help avert confrontations, build resilient communities, and foster fairness. Recovered ecosystems would help cope with climate shocks and preserve biodiversity. In a quantification of these benefits, policymakers rationalize remediation and design programs that will support the development of the larger socioeconomic construct.

To maximize the results of remediation, different policy instruments that may be used by the governments are pollution taxes and fees, subsidies and incentives, liability insurance and compensation funds, and public-private partnerships. Taxing polluters is an advantageous strategy that offers the appropriate motivation to prevent pollution, and it grants funds to cleanup programs. Remuneration for the implementation of clean technologies is motivators that induce industry compliance and innovation. Ensures populations impacted with insurance and restitution funds so that responsible parties fund cleanup and payment of damages. As a functional solution, they utilize collective resources and expertise to scale resounding remediation initiatives.

8.8 Challenges and Opportunities in Policy and Practice

8.8.1 Barriers to Effective Policy Implementation

Although acknowledgment is growing that there needs to be comprehensive HM remediation policies, there are many challenges to overcome in their implementation: Most low-and middle-income countries have financial and technical constraints on their resources and can hardly afford to enforce regulations and follow large-scale remedies. Such limits are compounded by the unavailability of trained staff, monitoring equipment, and laboratory facilities. The duplication of mandates by governmental agencies is inefficient and leads to confusion, a delay in action, and regulatory enforcement gaps. It is still challenging to coordinate various sectors that include the environment, health, and industry. There are instances where powerful polluters might seek excess political power to weaken the regulations or evade enforcement practices to continue with the pollution and cause distrust in people. Gaps in the data can be attributed to insufficient environmental monitoring and health surveillance systems to create dependable data to support decision-making. This impedes prioritization of sites, risk assessments, and tracking. Lack of community trust and engagement with authorities has their roots in the historical marginalization and broken promises, and contributes to the lack of social support in successful long-term remediation.

8.8.2 Emerging Trends and Innovations

In recent years, efforts have been made to overcome some of these hindrances in positive ways: bioremediation with plants or microbes, phytoremediation, and nanotechnology (NT) based solutions offer the relatively cheap and ecologically favorable cleanup of HMs, which can often be adapted to the local factors. Real-time access to environmental data on dedicated portals and mobile applications allows monitoring in real-time, giving access to communities and NGOs to contribute to oversight. Cross-sectoral solutions in which environmental, health, and economic governance are integrated are better able to ensure policy coherence, whereas adaptive governance can respond to environmental and economic changes and to new knowledge. In the transboundary pollution control, as well as resource-poor country capacity building, more and more are being funded by regional agreements and international funding programs.

8.9 Recommendations

Priorities should be established as part of an attempt to raise a potent remediation in terms of sustainability, equity, and effectiveness. Policies should be enacted in a counter-situational fashion that recognizes international best practice, yet is also sensitive to local socio-economic and cultural realities. The capacities of institutions should be enhanced by giving training, technical help, and better allocation of resources. Transparency and accountability should be improved and promoted citizen involvement by formalizing processes of community input and independent scrutiny. Innovation and transfer of technology should be facilitated to increase access to the right remediation technology. Trans-jurisdictional and cross-level governance arrangements should be supported that align agencies, stakeholders, and jurisdictions. The study should be encouraged, and the collection of data should be enhanced on patterns of contamination, ill effects, and the effectiveness of remediation.

8.10 Conclusion

Heavy metal (HM) pollution is a deep-seated and long-term environmental issue, which is socioeconomically and even politically significant. It has multidimensional impacts in terms of human health, nature-based ecosystems, and economic and social welfare. The undue burden on the backs of vulnerable and marginalized people highlights the fact that environmental and social justice go hand in hand in the development of effective remediation strategies. This study has spelled out how sustainable HM remediation is beyond technological fixes, and it needs to have an integrated strategy that welcomes inclusive governance, well-structured policy frameworks, and, not least, informed citizenry involvement. International conventions/national laws are vital resources, but they must be successfully enforced, open, and adapted to the context. The three-player (government, industry, and civil society) model or the four-player model (government, industry, civil society, and local communities) are sign of emergence bearing encouraging solutions of enhancing accountability and efficacy. CBA and social-economic estimation support the idea that the investments in remediation pay off enormously in terms of health, economic, and societal benefits, and as such, warrant the concentration of resources and political interest. Nevertheless, institutional fragmentation, resource constraints, political influence, and data deficiency are some of the problems that ought to be addressed systematically. Even in the future, it is necessary to incorporate the ideas of environmental justice into a guiding system to make sure that the remediation can foster not only environmental restoration but also human dignity. Most of the barriers can be overcome after adopting technological innovations, digital transparency, and multilateral collaboration. The overall vision is to integrate science, technology, and sustainability into social governance to restore damaged lands into more favorable and just landscapes. Future-proofing and fair remediation of HMs needs a commitment by all stakeholders to shared accountability, participatory dialogue, and resilient solutions, the basis of resilient communities and ecological systems into the future.

List of Abbreviations

As	Arsenic
Au	Gold
CABs	Community advisory boards

CBA	Cost-benefit analysis
Cd	Cadmium
CERCLA	Comprehensive Environmental Response, Compensation, and Liability Act
Hg	Mercury
HMs	Heavy metals
IQ	Intelligence quotient
NGOs	Non-Governmental Organizations
NT	Nanotechnology
Pb	Lead
PPP	Polluter-pays principle
US	United States

AI disclosure statement: The author confirms that no artificial intelligence (AI)-generated content or AI-assisted writing tools were used in the preparation of this chapter (including summaries and the abstract). All text, analysis, and conclusions are the result of the author's original work.

References

1. Alam, S., Bhardwaj, L.K., Mallick, R., Rai, S., Estimation of heavy metals and fluoride ion in vegetables grown nearby the stretch of River Yamuna, Delhi (NCR), India. *Indian J. Environ. Prot.*, 43, 1, 64–73, 2023.
2. Bhardwaj, L.K., Sharma, S., Jindal, T., Estimation of physico-chemical and heavy metals in the lakes of Grovnes & Broknes Peninsula, Larsemann Hill, East Antarctica. *Chem. Afr.*, 6, 5, 2677–2694, 2023.
3. Alloway, B.J., Heavy metals and metalloids as micronutrients for plants and animals, in: *Heavy Metals in Soils: Trace Metals and Metalloids in Soils and their Bioavailability*, pp. 195–209, Springer, Dordrecht, Netherlands, 2012.
4. Wu, X., Cobbina, S.J., Mao, G., Xu, H., Zhang, Z., Yang, L., A review of toxicity and mechanisms of individual and mixtures of heavy metals in the environment. *Environ. Sci. Pollut. Res.*, 23, 9, 244–8259, 2016.
5. Santa Maria, M.P., Hill, B.D., Kline, J., Lead (Pb) neurotoxicology and cognition. *Appl. Neuropsychol. Child*, 8, 3, 272–293, 2019.
6. World Health Organization (WHO), Report on activities during the ninth International lead poisoning prevention week, 24–30 October 2021, World Health Organization, 2023.
7. Järup, L. and Åkesson, A., Current status of cadmium as an environmental health problem. *Toxicol. Appl. Pharmacol.*, 238, 3, 201–208, 2009.
8. Grandjean, P., Weihe, P., White, R.F., Debes, F., Araki, S., Yokoyama, K., Murata, K., Sørensen, N., Dahl, R., Jørgensen, P.J., Cognitive deficit in

7-year-old children with prenatal exposure to methylmercury. *Neurotoxicol. Teratol.*, 19, 6, 417–428, 1997.
9. Smith, A.H., Lingas, E.O., Rahman, M., Contamination of drinking-water by arsenic in Bangladesh: a public health emergency. *Bull. World Health Organ.*, 78, 9, 1093–1103, 2000.
10. Naujokas, M.F., Anderson, B., Ahsan, H., Aposhian, H.V., Graziano, J.H., Thompson, C., Suk, W.A., The broad scope of health effects from chronic arsenic exposure: update on a worldwide public health problem. *Environ. Health Perspect.*, 121, 3, 295–302, 2013.
11. Boachie-Mensah, F.R.A.N.K., The Mining-Agriculture Nexus And Food Security In Kenyasi, Ahafo Region, Doctoral dissertation, University of Ghana, 2021.
12. Sharma, K., Rajan, S., Nayak, S.K., Water pollution: Primary sources and associated human health hazards with special emphasis on rural areas, in: *Water Resources Management for Rural Development*, pp. 3–14, Elsevier, Amsterdam, Netherlands, 2024.
13. Mamtani, R., Stern, P., Dawood, I., Cheema, S., Metals and disease: A global primary health care perspective. *J. Toxicol.*, 2011, 1, 319136, 2011.
14. Lanzi, E. and Dellink, R., Economic interactions between climate change and outdoor air pollution, 2019.
15. Ali, S.H., *Mining, the environment, and indigenous development conflicts*, University of Arizona Press, Tucson, Arizona, 2009.
16. Pulido, L., Flint, environmental racism, and racial capitalism. *Capital. Nat. Social.*, 27, 3, 1–16, 2016.
17. Hilson, G. and Pardie, S., Mercury: An agent of poverty in Ghana's small-scale gold-mining sector?. *Resour. Policy*, 31, 2, 106–116, 2006.
18. Bose-O'Reilly, S., McCarty, K.M., Steckling, N., Lettmeier, B., Mercury exposure and children's health. *Curr. Probl. Pediatr. Adolesc. Health Care*, 40, 8, 186–215, 2010.
19. Templeton, J. and Kohler, P., Implementation and compliance under the Minamata Convention on Mercury. *Rev. Eur. Comp. Int. Environ. Law*, 23, 2, 211–220, 2014.
20. Islam, M., The Basel Convention on the control of transboundary movements of hazardous wastes and their disposal: Critical analysis. *Int. J. Multidiscip. Res. Growth Eval.*, 1, 2, 11–16, 2020.
21. Chemicals, U.N.E.P., Minamata Convention on Mercury: text and annexes, 2025.
22. Grasso, D., Comprehensive environmental response, compensation, and liability act (CERCLA), in: *Hazardous Waste Site Remediation*, pp. 27–46, Taylor & Francis Group, Routledge, 2017.
23. Dietz, T., Ostrom, E., Stern, P.C., The struggle to govern the commons. *Science*, 302, 5652, 1907–1912, 2003.

9

The Effects of Heavy Metals from Dumpsite Leachate on Water Resources, Human Health Risks, and Management Strategies in Developing Countries

Demamu Tagele Haligamo[1]*, Tamru Tesseme Aragaw[1] and Esayas Alemayehu[2]

[1]*Faculty of Water Supply and Environmental Engineering, Water Technology Institute, Arba Minch University, Arba Minch, Ethiopia*
[2]*Faculty of Civil and Environmental Engineering, Jimma Institute of Technology, Jimma University, Jimma, Ethiopia*

Abstract
This chapter examined landfill characteristics in developing countries, highlighting that two-thirds are unsanitary open landfills, causing environmental and public health risks. Sanitary landfills with proper leachate and liner systems are rare but offer sustainable solutions. This chapter revealed high variability in electrical conductivity (EC), pH, total dissolved solids (TDS), chloride (Cl^-), and nitrate (NO_3^-) due to landfill age and waste composition. Most leachates showed suitability for biological treatment (biological oxygen demand (BOD)/chemical oxygen demand (COD) > 0.3, except for some landfills. The chapter identified high levels of iron (Fe), cadmium (Cd), and lead (Pb) in groundwater exceeding World Health Organization (WHO) limits, posing health risks. As observed, physical-chemical treatment methods showed heavy metal (HM) removal efficiency of > 95%, and biological methods achieved efficiency of 80–99%. Sludge production and management, environmental friendliness, technology costs, and scalability are factors for the selection of sustainable leachate treatment technologies. Two case studies indicated that there were severe environmental health risks caused by unsanitary landfills, but there is a potential for improving the landfills using

**Corresponding author*: demamutagele756@gmail.com; ORCID: https://orcid.org/0000-0002-3592-1024

Prangya Rath, Laxmi Kant Bhardwaj and Amit Kumar (eds.) Heavy Metal Remediation, (215–242)

engineered methods. This study suggests the construction of sanitary landfills and environmental monitoring to prevent pollution of water resources near landfills and to sustainably protect the environment and public health.

***Keywords*:** Heavy metals, health risks, landfills, leachate, leachate treatment, waste disposal, water pollution

9.1 Introduction

9.1.1 Characteristics of the Reviewed Landfill Sites

Both sanitary and open landfills were identified, but about two-thirds were unsanitary open landfills in low- and lower-middle-income countries. A landfill from Bangladesh was initially operated as an open landfill, but was finally modified to a semi-aerobic sanitary landfill with leachate collection and a liner system [1], and might serve as a good example to modify open landfills. All sanitary landfills except the one from Ghana had a leachate collection and liner system [2, 3], but all open landfills didn't have a leachate collection and liner system [4–6], which negatively affected the environment and nearby residents. Because of the negative effects from open landfills, complaints were raised by residents living mainly around them, which finally resulted in closing the sites, for example, an open landfill found in Côte d'Ivoire in West Africa that was closed in 2019 [4]. Ghana's landfills are mostly open dumps without leachate collection and treatment systems, situated in sensitive ecological or hydrological zones. As a result, the community is subjected to aesthetic burdens and public health hazards due to unsafe disposal facilities [7]. Almost all the reviewed open landfills were surrounded by residential houses [8, 9], but sanitary landfills were far away from the residential environment [1, 2, 10]. This study revealed that sanitary landfills occupied larger areas receiving a standard volume of waste when compared to the open landfills receiving greater volume of waste, for instance sanitary landfill holding 500 hectares from Ghana receives 432000 tons of waste per year [10], but the open landfill holding area of only 8 hectares from Nigeria receives a greater 810000 tons of waste per year [11] (Table 9.1).

9.1.2 Characterization of Landfill Leachate, Groundwater, and Surface Water Quality

In the reviewed investigations, the average TDS concentration of landfill leachates ranged from 783.0 to 2027.0 mg/L [9, 14] to maximum

Table 9.1 Characteristics of some of the reviewed landfills from developing countries.

S. no.	Landfill type, area, and waste volume it receives per year	Year it begins	Leachate collection and liner system	Distance from residents	Country	References
1	Open dumpsite, 153 hectares and 550000 tons	1965	No	Surrounded by residents	Côte d'Ivoire	[4]
2	Sanitary landfill, 21 hectares and 66600 tons	2017	Yes	500 m away from residents	Bangladesh	[1]
3	Sanitary landfill, 194 hectares and 756000 tons	2001		It's away from residents	Egypt	[2]
4	Sanitary landfill, 40.5 hectares and 270000 tons	2003		At a certain distance	India	[3]
5	Open landfill, 34.4 hectares, and unknown waste volume	2004	No	Residents within 500 m		[12]

(*Continued*)

Table 9.1 Characteristics of some of the reviewed landfills from developing countries. (*Continued*)

S. no.	Landfill type, area, and waste volume it receives per year	Year it begins	Leachate collection and liner system	Distance from residents	Country	References
6	Sanitary landfill, 40 hectares, and unknown waste volume	1992	Yes	Surrounded by residents	Nigeria	[13]
7	Open landfill, 8 hectares and 810000 tons	2008	No	Bordered by a resident's house		[11]
8	Sanitary landfill, 500 hectares and 432000 tons	2003		Far from a residential area	Ghana	[10]
9	Open dumpsite, 3–4 hectares, and unknown waste volume	2006		Close to residents	Ethiopia	[8]
10	Open landfill, 37 hectares and 470000 tons	1964		Residents within 40 m		[5]
11	Open landfill, 12 hectares and 43200 tons	2001		Surrounded by residents	Iran	[6]

concentrations of 27452.0–62000.0 mg/L [2, 11]. Variation in the age and waste composition, as well as the availability of leachate collection and liner systems, may be the cause of the variations in TDS concentrations among landfill leachates of the reviewed studies [15]. TDS concentrations in surface and ground waters close to landfills varied between 92.0 and 1447.0 mg/L [14, 16], and 56.0 and 9895.0 mg/L [2, 17], respectively. The reviewed studies' landfill leachates had mean EC values ranging from 68.0 to 40921.0 μS/cm. The study identified low EC values of 68.0 and 391.0 μS/cm [9, 18] and high values of 40921.0 and 5275.0 μS/cm [2, 4] in landfill leachate. The high value of EC in water indicates an abundance of anions and cations [1]. The average EC value in surface and ground water from the reviewed studies ranged from 148.0 to 7181.0 μS/cm [4, 17], and from 62.0 to 12745.0 μS/cm [2, 4], respectively.

Water acidity and alkalinity are indicated by its pH. This study identified that the pH in leachate ranged from 6.0 to 8.8 [12, 18]. Studies showed that the pH of landfill leachate can be used for categorizing landfills. Landfills classified as young have a pH scale of < 6.5 and have been in operation for 0–10 years, intermediate landfills have a pH scale from 6.5 to 7.5 and have been in operation for 10–20 years, and old landfills have a pH > 7.5 and have been in operation for over 20 years [19, 20]. As a result, in this study a young landfill was identified as a dumping site from Bangladesh [18], while four had reviewed dumping sites as intermediate landfills [2, 14, 16, 21], and the old landfills comprised of other seven dumping sites [1, 4, 5, 9, 11, 12, 17]. The pH scales in nearby surface water and groundwater were varied between 6.4–8.4 [10, 14] and 5.3–7.9 [2, 4], respectively.

The mean Cl^- concentration in the reviewed landfill leachate varies from 64.0 mg/L [18] to 11387.0 mg/L [2]. Concentration of Cl^- in samples of water was observed to range from 2.3 to 308.0 mg/L [5, 18] in surface water and from 4.0 to 6890.0 mg/L [2, 18] in ground water. Concentration of NO_3^- in landfill leachate was observed to vary between 1.4 mg/L [2] and 395.0 mg/L [3]. Similarly, NO_3^- concentration in samples of surface water as well as groundwater were reported to vary in a range between 0.04–38.5 mg/L [6, 8] and 0.25–68.3 mg/L [2, 12], respectively. BOD concentrations in landfill leachates were observed to range between 383.0 and 16693.0 mg/L [16, 18], while COD concentrations were reported to range between 160.0 and 33824.0 mg/L [11, 16]. The reviewed studies revealed that concentrations of BOD and COD in surface water ranged between 2.0–105.0 mg/L and 4.1–1080.0 mg/L [14, 16], respectively. BOD and COD concentrations in the reviewed groundwater varied between 0.0–60.0 mg/L [2, 21] and 0.0–440.0 mg/L [14, 21], respectively. BOD to COD ratio can be taken as a benchmark to determine the suitability of biological techniques

Table 9.2 Ratio of biological oxygen demand (BOD) to chemical oxygen demand (COD) to check the applicability of biological processes in leachate treatment.

S. no.	Parameter	Ratio	References
1	Biological Oxygen Demand (BOD)/ Chemical Oxygen Demand (COD)	0.69	[2]
		0.70	[3]
		0.16	[12]
		0.64	[18]
		0.66	[9]
		0.14	[14]
		0.50	[5]
		0.49	[16]
		0.32	[21]

for landfill leachate treatment. Biological processes are effective in organic matter removal from leachates when the ratio of BOD/COD is >0.3, but a lower BOD/COD ratio indicates low removal performance and inability to use the biological process [22, 23]. All of the BOD/COD ratios showed that biological processes are appropriate and effective to remove organic matter from landfill leachates, except for the findings of two Indian studies [12, 14] (Table 9.2).

9.1.3 Heavy Metals (HMs)

The six HMs that have been carefully reviewed and discussed include iron (Fe), zinc (Zn), cadmium (Cd), chromium (Cr), lead (Pb), and nickel (Ni).

9.1.3.1 Iron (Fe) Concentration

Reviewed studies have reported that the average concentration of Fe in landfill leachate was found between 0.21 mg/L [21] and 1473.0 mg/L [13] (Table 9.3). Fe concentration in surface water as well as groundwater was varied between 0.16 mg/L [14] to 1.66 mg/L [18], and 0.04 mg/L [6] to 2.29 mg/L [24], respectively. Based on the WHO standard limit, Fe concentration in drinking water could be below 0.1 mg/L [25]. The review

Table 9.3 Quality of leachate from dumping sites in developing countries.

S. no.	Parameters							References
	Iron (Fe)	**Copper (Cu)**	**Zinc (Zn)**	**Cadmium (Cd)**	**Chromium (Cr)**	**Lead (Pb)**	**Nickel (Ni)**	
1	6.45 ± 1.88	0.40	0.515 ± 0.005	0.110 ±0.020	NS	0.550 ± 0.010	0.315 ± 0.005	[4]
2	NS			0.003	0.018	0.050	0.16	[1]
3	6.32± 1.83	0.07 ± 0.03	0.749 ± 0.235	0.094 ± 0.026	0.062 ± 0.044	0.019 ± 0.004	0.096 ± 0.015	[2]
4	5.40 ± 2.61	0.33 ± 0.31	0.000		1.770 ± 0.350	0.000		[12]
5	1473.16 ± 413.59	NS	23.880 ± 14.400	7.120 ± 4.800	1.500 ± 0.910	3.000 ± 1.040	2.420 ± 2.310	[13]
6	5.93 ± 1.32	0.39 ± 0.26	1.410 ± 1.10	0.070 ± 0.020	1.259 ± 0.740	0.020 ± 0.007	NS	[18]
7	6.50	NS	9.000	8.800	NS	10.200		[11]
8	NS	0.26 ± 1.08	0.540 ± 0.200	0.300 ± 0.010		0.080 ± 0.100	0.400 ± 0.100	[9]

(Continued)

Table 9.3 Quality of leachate from dumping sites in developing countries. (*Continued*)

S. no.	Parameters							References
	Iron (Fe)	Copper (Cu)	Zinc (Zn)	Cadmium (Cd)	Chromium (Cr)	Lead (Pb)	Nickel (Ni)	
9	11.16	0.15	3.000	0.035	0.021	0.300	1.339	[14]
10	NS		3.260	0.140	NS	0.770	NS	[17]
11	25.61 ± 2.85	1.31 ± 1.08	6.092 ± 3.885	1.083 ± 0.065	1.918 ± 1.900	3.574 ± 2.228		[24]
12	6.59 ± 4.05	0.36 ± 0.12	0.200 ± 0.140	NS	0.430 ± 0.380	0.180 ± 0.110	0.690 ± 0.460	[16]
13	0.21	NS	0.084		0.000	0.035	0.017	[21]

found that, with the exception of one study that indicated a Fe concentration of 0.04 mg/L in groundwater, the concentration of Fe in surface water and groundwater close to dumping sites was above the statutory limit [6] (Tables 9.4 and 9.5).

9.1.3.2 Zinc (Zn) Concentration

According to the evaluated studies, the content of Zn in landfill leachate varied from 0.00 mg/L [12] to 23.88 mg/L [13] (Table 9.3). Around the dumping sites, Zn concentrations in the groundwater and surface water ranged from 0.00 mg/L [6] to 3.83 mg/L [6] to 3.83 mg/L [17], and 0.00 mg/L [6] to 1.57 mg/L [17], respectively. According to the WHO guideline limit, drinking water should have a Zn concentration of less than 3.0 mg/L [25]. With the exception of one study's reported Zn value of 3.83 mg/L in surface water, reviewed research showed that both groundwater and surface water had Zn concentrations within the WHO guideline limit [17] (Tables 9.4 and 9.5).

9.1.3.3 Cadmium (Cd) Concentration

According to reviewed research, the amount of Cd in landfill leachate ranged from 0.00 mg/L [12] to 8.80 mg/L [11] (Table 9.3). Surface water and groundwater surrounding dumping sites were found to have average Cd concentrations of 0.00 mg/L [9] and 0.09 mg/L [4], and 0.00 mg/L [11] and 0.16 mg/L [10], respectively. The WHO standard recommendation states that the amount of Cd in drinking water should not exceed 0.003 mg/L [25]. Nearly half of the reviewed studies reported Cd concentration in surface and ground water within the WHO standard limit. A study from India identified HMs, including Pb, copper (Cu), Cd, and chromium (Cr), from different studied sites [26]. Higher concentrations of Cd in drinking water can cause renal, arterial hypertension, cramps, diarrhea, nausea, and vomiting [27]. Nearly half of the surface and ground waters close to landfills indicated higher levels of Cd from Côte d'Ivoire, Egypt, Bangladesh, Ghana, and Nigeria that might cause health risks (Tables 9.4 and 9.5).

9.1.3.4 Chromium (Cr) Concentration

According to the evaluated research, the content of Cr in landfill leachate ranged from 0.000 mg/L [21] to 1.918 mg/L [24] (Table 9.3). The levels of Cr in groundwater and surface water close to disposal sites varied from 0.000 mg/L [6] to 0.362 mg/L [18], and 0.000 mg/L [14] to 0.760 mg/L [18].

Table 9.4 Groundwater quality near dumping sites in developing countries.

S. No.	Parameters							References
	Iron (Fe)	**Copper (Cu)**	**Zinc (Zn)**	**Cadmium (Cd)**	**Chromium (Cr)**	**Lead (Pb)**	**Nickel (Ni)**	
1	0.16 ± 0.16	0.62 ± 0.42	0.030 ± 0.001	0.046 ± 0.015	NS	0.633 ± 0.131	0.057 ± 0.006	[4]
2	NS			0.002	0.0052	0.003	0.046	[1]
3	1.23 ± 0.74	0.03 ± 0.02	0.148 ± 0.032	0.011 ± 0.006	0.039 ± 0.015	0.005 ± 0.001	0.057 ± 0.020	[2]
4	1.06 ± 4.05	NS						[12]
5	1.25 ± 0.44	0.04 ± 0.02	0.028 ± 0.018	0.013 ± 0.008	0.362 ± 0.300	0.002 ± 0.001	NS	[18]
6	0.57 ± 0.34	NS	0.008 ± 0.004	0.000	NS	0.000		[11]
7	0.25 ± 0.15	0.00	NS	0.160 ± 0.140				[10]

(Continued)

Table 9.4 Groundwater quality near dumping sites in developing countries. (*Continued*)

S. no.	Parameters							References
	Iron (Fe)	**Copper (Cu)**	**Zinc (Zn)**	**Cadmium (Cd)**	**Chromium (Cr)**	**Lead (Pb)**	**Nickel (Ni)**	
8	0.62	0.00	0.400	0.000	0.000		0.000	[14]
9	NS		1.570		NS		NS	[17]
10	2.29 ± 1.76	0.25 ± 0.02	0.237 ± 0.076	0.033 ± 0.004	0.039 ± 0.023	0.011 ± 0.008		[24]
11	0.04 ± 0.04	NS	0.000			0.0002 ± 0.00037	0.0126	[6]
12	NS		0.018	NS	0.000	0.027	0.002	[21]

Table 9.5 Surface water quality near dumping sites in developing countries.

S. no.	Parameters							References
	Iron (Fe)	Copper (Cu)	Zinc (Zn)	Cadmium (Cd)	Chromium (Cr)	Lead (Pb)	Nickel (Ni)	
1	0.84 ± 0.28	0.208 ± 0.195	0.041± 0.015	0.090 ± 0.015	NS	0.584 ± 0.300	0.088 ± 0.055	[4]
2	NS			0.002	0.044	0.010	0.053	[1]
3	0.68 ± 0.84	NS	0.010 ± 0.005	0.020 ± 0.010	0.100 ± 0.110	0.260 ± 0.240	0.000	[13]
4	1.66 ±1.26	0.106 ± 0.015	0.046 ± 0.026	0.054 ± 0.009	0.760 ±0.310	0.003 ±0.002	NS	[18]
5	NS	0.020 ± 0.950	0.390 ± 0.180	0.000	NS	0.000	0.080 ± 0.100	[9]

(Continued)

Table 9.5 Surface water quality near dumping sites in developing countries. (*Continued*)

S. no.	Parameters							References
	Iron (Fe)	Copper (Cu)	Zinc (Zn)	Cadmium (Cd)	Chromium (Cr)	Lead (Pb)	Nickel (Ni)	
6	0.35 ± 0.25	0.000	NS	0.035 ± 0.017			NS	[10]
7	0.16		1.000	0.000	0.000		0.000	[14]
8	NS		3.830		NS		NS	[17]
9	0.45 ± 0.17	0.100 ± 0.010	0.360 ± 0.090	NS	0.060 ± 0.020		0.020 ± 0.010	[16]
10	1.10 ± 0.67	NS	0.000		0.008 ± 0.003	0.0023 ± 0.0022	0.029 ± 0.011	[6]
11	NS		0.014	NS	0.000	0.024	0.006	[21]

According to the WHO standard guideline, Cr concentration in water should be below 0.05 mg/L [25]. The Cr concentration of surface waters in nearly half of the reviewed studies was within the WHO standard limit (Table 9.5), and the Cr concentration in groundwater was within the standard limit in all reviewed studies except the one that reported 0.362 mg/L [18] (Table 9.4).

9.1.3.5 Lead (Pb) Concentration

Lead (Pb) is a toxic HM that pollutes the ecosystem from human sources like landfill leachate [28]. Pb concentrations from landfill leachate ranged from 0.00 mg/L [12] to 10.20 mg/L [11] according to reviewed studies (Table 9.3). In surface water, the Pb concentration was 0.000 mg/L [9] and 0.584 mg/L [4], whereas in groundwater, it was 0.000 mg/L [11] and 0.633 mg/L [4]. The WHO standard limit states that drinking water should have a Pb concentration of less than 0.01 mg/L [25]. Pb concentrations of surface waters of the reviewed studies were within the WHO standard limit, except for three studies [4, 13, 21] (Table 9.5). Likewise, for groundwater, all were within the standard limit except for three studies [4, 21, 24] (Table 9.4). Pb can harm the brain of children and disrupt the nervous system if the content in drinking water is greater than 0.01 mg/L [29]. Six surface and ground waters from Ghana, Côte d'Ivoire, Nigeria, and India that are located close to landfills with Pb concentrations exceeding 0.01 mg/L were recognized in this study as having a high likelihood of posing a major risk to human health when consumed (Tables 9.4 and 9.5).

9.1.3.6 Nickel (Ni) Concentration

The examined studies' average Ni concentrations in landfill leachate varied from 0.00 mg/L [12] to 2.42 mg/L [13] (Table 9.3). Surface and ground water were found to have average Ni concentrations of 0.000 mg/L [14] and 0.088 mg/L [4], and 0.000 mg/L [14] and 0.057 mg/L [2], respectively. According to the WHO standard guideline, Ni concentration in water should be below 0.07 mg/L [29]. The Ni concentration of surface water in reviewed studies was within the WHO standard limit, except for two studies [4, 9] (Table 9.5), and the Ni concentration in groundwater was within the standard limit in all the reviewed studies (Table 9.4).

9.1.4 Objectives of the Critical Review

There were four primary areas of focus for the reviews. It began by analyzing current evidence about the dumping sites' characteristics and characterizing water resources and leachate quality. Then it explored the effects of the landfill leachate on the local water resources and associated health risks. After that, it identified the obstacles and challenges in waste disposal practice and existing and emerging treatment technologies. Finally, a brief conclusion about the review result and key recommendations was provided. The study provided valuable inputs for municipalities, policymakers, and researchers to establish well-designed sanitary landfills, amend and enforce existing rules, and carry out additional research to find further solutions.

9.2 Review of Literature

A qualitative purposive research approach was employed in the selection of the reviewed publications. To obtain reliable information from a wide range of sources, this review looked at different web searches. The majority of the reviews (88%) were conducted using publications from the last 15 years and written in English. The entire review process is illustrated in Figure 9.1. Many publications were found by searching for keywords. Nearly 70% of the selected publications were from Scopus-indexed journals. The selection process was further modified by manual screening, which gave priority to publications addressing developing countries. During the procedure, publications were extracted for further analysis. At the same time, EndNote software was applied to manually screen the abstracts and conclusions of suggested pertinent papers. Under this study, selection criteria were applied to reduce the downloaded publications to the interest and purpose of the review. The selection criteria employed for the publications included the focus areas and years of publication. Finally, about 55 publications directly related to the title were found relevant to the review subject, then downloaded and reviewed.

9.3 Public Health Risks of Groundwater Near Landfills

Concerning health risks related to groundwater quality, the average concentration of reviewed contaminants exceeding the WHO standard limit

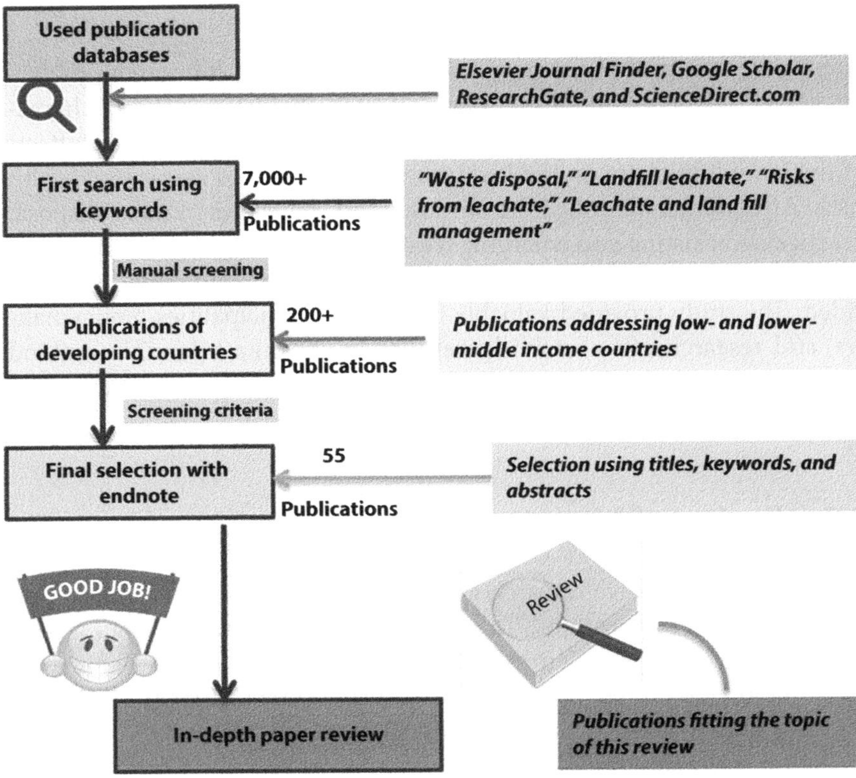

Figure 9.1 The framework employed for the literature review.

was discussed [25]. Eight reviewed articles had average Fe concentrations in groundwater near landfills that were higher than the recommended level, which could result in skin damage, gastrointestinal problems, and hemochromatosis. According to the five reviewed studies, the average Cd concentrations in groundwater were likewise higher than the permissible limit, which can lead to issues with palatability, indigestion, and taste. The mean TDS concentrations in the four reviewed studies were above the WHO standard limit, which can decrease water palatability and cause irritation of the gastrointestinal system. The average Pb concentrations of the three examined studies were higher than the advised threshold, which may cause toxicity to reproductive organs, hypertension, kidney impairment, and brain damage (Tables 9.4 and 9.6).

Table 9.6 Groundwater quality parameters that exceeded the World Health Organization (WHO) limit and associated health risks.

S. no.	Parameters	World Health Organization (WHO) limit [25]	Studies reported beyond the World Health Organization (WHO) limit (refer to Table 9.4)	Related health effects if it exceeds the maximum level
1	Chloride (Cl^-) mg/L	250.0	1 study	Affects taste, indigestion, and palatability.
2	Nitrate (NO_3^-) mg/L	10.0	2 studies	Cause methemoglobinemia, increasing heart rate, and abdominal cramps.
3	Total dissolved solids (TDS) mg/L	1000.0	4 studies	Decrease palatability and cause gastrointestinal irritation.
4	pH	6.2–8.8	2 studies	The effects include taste, causing kidney disease, liver disease, stomach cramps, and diarrhea.
5	Iron (Fe) mg/L	0.1	8 studies	Causes hemochromatosis, stomach problems, nausea, and damaged skin.
6	Lead (Pb) mg/L	0.01	3 studies	It leads to hypertension, kidney impairment, brain damage, and toxicity to reproductive organs.
7	Cadmium (Cd) mg/L	0.003	5 studies	It causes kidney disease, fragile bones, and cancer.
8	Chromium (Cr) mg/L	0.05	1 study	It causes liver damage, nerve tissue damage, kidney problems, and skin rashes.

9.4 Challenges and Obstacles to Waste Disposal in Developing Countries

The following challenges and barriers have been cited as causes of improper waste disposal in developing nations.

9.4.1 Institutional Capability, Functions, and Responsibility

According to a study, local waste management authorities lack professional expertise and organizational leadership styles [30]. A report revealed that it is difficult to understand the complicated challenges of solid waste disposal since very little information is available in the different sectors involved [31]. The majority of urban cities in developing countries lack experience with managed disposal. Though only a few were actively managed and had disposal places. Waste dumping in open spaces and along roadsides is more likely when there is a shortage of waste containers and a greater distance to them [30].

9.4.2 Political Dedication and Sociocultural Elements

Political leaders in developing countries rarely give solid waste disposal top priority unless there is a significant and engaged public interest. As a result, there is little motivation to supervise unauthorized practices or invest in disposal [7]. Most people believe that local governments are the only ones responsible for managing waste, and that the general public is not supposed to help. For this reason, scholars mentioned sociocultural aspects such as decision-making participation, community awareness, and societal apathy towards contributing to solutions [32]. Politicians prioritize other municipal tasks over solid waste management (SWM), which leads to a shortage of experienced waste experts in municipalities [33].

9.4.3 Financial Factors and Technical Standards

Prioritizing a sanitary landfill over others requires resources and can be challenging [34]. Short-term funding may be available for the construction of a disposal site, but consistent income is required for ongoing expenses over the long run. If residents refuse to pay for waste disposal, it is extremely difficult to offer the service. The safe removal of trash in well-designed and engineered landfills is restricted by a lack of funding [30, 35]. The delivery of appropriate waste disposal services also depends on the lack of financial support, the scarcity of resources [35], and a lack of proper use of economic instruments [33]. Technical factors affecting the disposal system include inadequate infrastructure [32], poor roads and vehicles, lack of technology and trustworthy data, and lack of technical skills among government officials and municipality experts [7, 36].

9.4.4 Capacity of Waste Recycling and Treatment Facility

Waste minimization and the use of 3Rs (reduce, reuse, and recycle) are cited as top-prioritized and effective long-term waste management strategies [37]. However, the actual waste treated at a waste incinerator and power-generating plant that functions as an intermediate treatment facility is far less than its capacity, and also, safety and operation-related problems have been identified. Because of this, only a small percentage of waste can be recycled [38], directly affecting the final waste disposal in developing nations.

9.4.5 Legislation and Legal Aspects

Another obstacle to the safe disposal of solid waste is the absence of legislation [30]. According to research, the expansion of an integrated waste management system is positively impacted by the legal framework [39], whereas it is negatively impacted by inadequate regulations and a lack of satisfactory policies [36], and weak regulations [30].

9.5 Technologies for Heavy Metal (HM) Removal from Landfill Leachate

The study discussed some existing and emerging technologies for HM removal, helping to choose the best method for HM removal to meet standards and minimize environmental damage.

9.5.1 Precipitation

Precipitation is one of the methods applied for the removal of HMs from wastewater/leachate using agents such as hydroxides, sulfides, and phosphates. Studies reported that precipitation achieved a removal efficiency of 97% for HMs [40, 41]. The effectiveness and efficiency of precipitation for metal removal depend on factors such as temperature, pH, and precipitant type, including ferric chloride and lime [42]. Major disadvantages of precipitation are sludge generation and increasing disposal costs. An increased treatment efficiency was observed when precipitation was combined with methods like ion exchange/nano-filtration [43].

9.5.2 Adsorption

Another mechanism for HM removal is adsorption. It removes HMs from wastewater/leachate through which pollutants can bind to solid adsorbents like activated carbon/clay/bio-materials. Adsorption showed a removal efficiency of 99% for metals such as Pb, Cd, and Cr. The effectiveness of adsorption depends on temperature, pH, and adsorbent type [44, 45]. Activated carbon from pecan shells made 98.6% Fe removal, while modified *Caryota urens* seeds removed 80.91 mg/g of Zn [46]. Low-cost alternatives such as industrial byproducts (Fe slag and sugarcane bagasse ash) and biochar showed promise, with over 90% removal efficiency [43].

9.5.3 Ion Exchange

By exchanging ions in solution with those bonded to synthetic or natural resins, ion exchange is a successful technique for eliminating HMs from wastewater. High removal efficiency of over 99.9% for metals (like Pb, Cd, and silver (Ag)) [47] employing materials including polyelectrolyte-modified membranes, zeolites, and chitosan-coated nanoparticles (NPs). Key factors such as pH, temperature, and resin selectivity typically describe the process [48, 49]. Iron oxide (Fe_3O_4)/chitosan NPs achieved 79.24 mg/g Pb (II) uptake, and synthetic zeolite showed exceptional capacity for Ag^+ (123.05 mg/g) and Pb^{2+} (227.70 mg/g) [48].

9.5.4 Phytoremediation

Phytoremediation is a treatment method that utilizes plants and their associated microorganisms to extract/stabilize/degrade HMs from contaminated water or wastewater. Phytoremediation is defined as a plant's ability to decrease pollution concentration in an environment by absorbing it through its roots, and accumulating or translocating it to a less toxic level in its various parts [50]. It is a cost-effective and eco-friendly treatment method compared to conventional physicochemical decontamination treatments [47, 50]. Aquatic plants like water hyacinth and duckweed showed significant metal uptake, with removal rates of 24–80% for Cu, Cr, and Pb due to their dense root systems and rapid growth [51]. Terrestrial plants such as African grass (*Pennisetum clandestinum*) also show resilience, tolerating landfill leachate for 180 days while removing Cr, Zn, and Ni [47]. The efficiency varies with metal speciation; Cr (III) accumulates more in roots than Cr (VI) [52].

9.5.5 Artificial Wetlands

Artificial wetlands are a natural treatment process to remove HM contamination, providing low-energy and sustainable remediation outcomes. It is less than 1.0 m deep and uses plants to achieve a removal efficiency of >99% for HMs (manganese (Mn), Pb, and Cu) when used in combination with a substrate like shellfish shells [53]. Mangrove species indicated superior resilience and removal rates up to 99.8% for Pb compared to non-mangrove plants [54]. The performance varies with design: horizontal wetlands outperformed vertical ones in Cr removal (92.1% vs. 35.4%), attributed to longer hydraulic retention and better plant-substrate interactions [55].

9.6 Case Studies from Two Different Countries

- **Case Study 1**

A dumping site found in Savar near Dhaka, called Aminbazar landfill, was initially an open landfill and was finally modified to a semi-aerobic sanitary landfill in 2017 to solve Dhaka's increasing waste problems. It covers about 21 hectares and is designed to process about 666000 metric tons of waste. The landfill is characterized by its leachate collection and treatment system to prevent the contamination of nearby groundwater sources. The landfill is managed by the Dhaka North City Corporation. It utilizes a cell-based waste disposal system with soil cover applied daily to avoid related environmental impact. The landfill utilizes a leachate treatment system (aeration, sedimentation, and biological processes) to avoid pollutants and to ensure safe discharge. Inconsistent maintenance and funding shortages have occasionally compromised efficiency, leading to temporary leachate seepage incidents [1].

- **Case Study 2**

São João landfill is Brazil's largest dumping site and is found in São Paulo, causing serious groundwater contamination owing to a lack of an adequate liner system. High concentrations of HMs (Pb and Cd), organic pollutants, and ammonia in the surrounding wells were identified, posing serious health risks to the nearby residents. To avoid contamination of the groundwater, concerned authorities implemented a treatment system through which contaminated groundwater was extracted and treated using biological reactors called activated sludge, which was then followed by reverse osmosis (RO). Furthermore, constructed wetlands were applied to reduce nitrogen concentration *via* phytoremediation. The RO system effectively removed above 90% of HMs and dissolved organic pollutants, while wetlands reduced ammonia concentration. After treatment, post-treatment monitoring was done, indicating a significant improvement in groundwater quality [56, 57].

9.7 Recommendations

Municipalities and other concerned bodies should construct well-engineered sanitary landfills or modify existing open landfills into sanitary landfills by constructing leachate collection and treatment systems, liner systems, and applying monitoring systems to protect the environment and human health, following the observed successful models taken from Bangladesh's modified landfill. Experts should prioritize biological treatment methods for landfill leachate, having determined BOD/COD ratios of >0.3, and select advanced physicochemical treatment methods/combined for mature landfills with low biodegradability. Concerned bodies should implement monitoring and remediation for landfill leachate to prevent HM pollution, specifically in urban areas with determined HM levels beyond the recommended standards. Municipalities and concerned sectors should strengthen their institutional and technical capacity *via* training, logistics improvement, and applying waste management standards to enhance safe, effective, and sustainable practices. There is a need to implement policies and financial reforms (waste service cost recovery and incentives for recycling) to ensure the sustained viability of waste disposal systems and reduce dependence on unsanitary open dumping. Environmental engineers and waste experts should combine highly efficient physicochemical methods with low-cost biological methods for balancing operations and sustainability to minimize the generation of sludge and energy utilization.

9.8 Conclusion

The conversion of existing open landfills into well-designed sanitary landfills with appropriate leachate collection and treatment systems is crucial to improving and maintaining waste disposal practices in developing nations, where open landfills are widely used methods that pose serious environmental and public health risks due to poor planning, construction, and increased urban waste volumes. The release of poorly managed landfill leachate was mainly observed in urban areas of developing countries, posing severe environmental risks to nearby water resources and public health risks to the surrounding residents. Landfills provide severe health problems since they contribute significantly to the pollution of nearby water resources with HMs. To safeguard local water supplies and public health, an efficient landfill leachate collection and treatment system is required. Sociocultural apathy, a lack of institutional capacity, financial limitations, a lack of technical standards, political negligence, a lack of recycling technologies, and a lack of legal frameworks were among the waste disposal challenges and obstacles found in developing nations. These factors exacerbate unsafe disposal practices and call for comprehensive and long-term solutions. Therefore, in order to transition from unsafe dumping to ecologically friendly and protective waste disposal systems, sustainable municipal solid waste disposal practices in developing nations necessitate coordinated efforts across administration, financing, technologies, and public participation.

List of Abbreviations

3Rs	Reduce, reuse, and recycle
Ag	Silver
BOD	Biochemical Oxygen Demand
Cd	Cadmium
Cl^-	Chloride
COD	Chemical Oxygen Demand
Cr	Chromium
Cu	Copper
EC	Electrical Conductivity
Fe	Iron
Fe_3O_4	Iron oxide
HM	Heavy metal
Mn	Manganese
Ni	Nickel
NO_3^-	Nitrate
NPs	Nanoparticles
NS	Not studied
Pb	Lead

RO	Reverse Osmosis
SWM	Solid waste management
TDS	Total Dissolved Solids
WHO	World Health Organization
Zn	Zinc

AI disclosure statement: The author acknowledges the use of AI-based tools for minor language editing or grammatical correction. However, all intellectual contributions, analyses, and conclusions presented in this chapter are entirely the author's own.

References

1. Hredoy, R.H., Siddique, M.A.B., Akbor, M.A., Shaikh, M.A.A., Rahman, M.M., Impacts of landfill leachate on the surrounding environment: A case study on Amin Bazar Landfill, Dhaka (Bangladesh). *Soil Syst.*, 6, 4, 90, 2022.
2. Abd El-Salam, M.M. and Abu-Zuid, G.I., Impact of landfill leachate on the groundwater quality: A case study in Egypt. *J. Adv. Res.*, 6, 4, 579–586, 2015.
3. Mor, S., Ravindra, K., Dahiya, R.P., Chandra, A., Leachate characterization and assessment of groundwater pollution near municipal solid waste landfill site. *Environ. Monit. Assess.*, 118, 1, 435–456, 2006.
4. Naminata, S., Kwa-Koffi, K.E., Marcel, K.A., Marcellin, Y.K., Assessment and impact of leachate generated by the Landfill City in Abidjan on the quality of ground water and surface water (M'Badon Bay, Côte d'Ivoire). *J. Water Resour. Prot.*, 10, 1, 145–165, 2018.
5. Haile, T. and Abiye, T.A., Environmental impact and vulnerability of the surface and ground water system from municipal solid waste disposal site: Koshe, Addis Ababa. *Environ. Earth Sci.*, 67, 1, 71–80, 2012.
6. Javahershenas, M., Nabizadeh, R., Alimohammadi, M., Mahvi, A.H., The effects of Lahijan landfill leachate on the quality of surface and groundwater resources. *Int. J. Environ. Anal. Chem.*, 102, 2, 558–574, 2022.
7. Jagun, Z.T., Daud, D., Ajayi, O.M., Samsudin, S., Jubril, A.J., Rahman, M.S.A., Waste management practices in developing countries: a socio-economic perspective. *Environ. Sci. Pollut. Res.*, 30, 55, 116644–116655, 2023.
8. Feven, T., Assessment of the Effect of Waste Dump Site in the Surrounding Surface and Ground Water Quality the Case of Bahirdar, Ethiopia, Doctoral dissertation, 2021.
9. Mekonnen, B., Haddis, A., Zeine, W., Assessment of the effect of solid waste dump site on surrounding soil and river water quality in Tepi town, Southwest Ethiopia. *J. Environ. Public Health*, 2020, 1, 5157046, 2020.

10. Amano, K.O.A., Danso-Boateng, E., Adom, E., Kwame Nkansah, D., Amoamah, E.S., Appiah-Danquah, E., Effect of waste landfill site on surface and ground water drinking quality. *Water Environ. J.*, 35, 2, 715–729, 2021.
11. Aderemi, A.O., Oriaku, A.V., Adewumi, G.A., Otitoloju, A.A., Assessment of groundwater contamination by leachate near a municipal solid waste landfill. *Afr. J. Environ. Sci. Technol.*, 5, 11, 933–940, 2011.
12. Mishra, S., Tiwary, D., Ohri, A., Agnihotri, A.K., Impact of Municipal Solid Waste Landfill leachate on groundwater quality in Varanasi, India. *Groundwater Sustainable Dev.*, 9, 100230, 2019.
13. Essien, J.P., Ikpe, D.I., Inam, E.D., Okon, A.O., Ebong, G.A., Benson, N.U., Occurrence and spatial distribution of heavy metals in landfill leachates and impacted freshwater ecosystem: An environmental and human health threat. *PLoS One*, 17, 2, e0263279, 2022.
14. Naveen, B.P., Sumalatha, J., Malik, R.K., A study on contamination of ground and surface water bodies by leachate leakage from a landfill in Bangalore, India. *Int. J. Geo-Eng.*, 9, 1, 27, 2018.
15. Hussieny, M.A., Morsy, M.S., Ahmed, M., Elagroudy, S., Abdelrazik, M.H., Municipal solid waste and leachate characterization in the Cairo Metropolitan Area. *Resources*, 11, 11, 102, 2022.
16. Tesseme, A.T., Vinti, G., Vaccari, M., Pollution potential of dumping sites on surface water quality in Ethiopia using leachate and comprehensive pollution indices. *Environ. Monit. Assess.*, 194, 8, 545, 2022.
17. Nyandwaro, E.O., The Impacts Of Solid Waste On Ground And Surface Water Quality In Kisii Municipality, Kenya, Unpublished Masters' Thesis, Environmental Science Department, School of Environmental Studies, Kenyatta University, Nairobi, Kenya. 2017.
18. Hossain, M.L., Das, S.R., Hossain, M.K., Impact of landfill leachate on surface and ground water quality, 2014.
19. Tejera, J., Miranda, R., Hermosilla, D., Urra, I., Negro, C., Blanco, Á., Treatment of a mature landfill leachate: comparison between homogeneous and heterogeneous photo-Fenton with different pretreatments. *Water*, 11, 9, 1849, 2019.
20. Tchobanoglous, G., Theisen, H., Vigil, S., *Integrated Solid Waste Management, 'Engineering Principles and Management Issues'*, McGraw–Hill, Inc., New York, p. 949, 1993.
21. Sharma, D. and Ganguly, R., Parametric analysis of leachate and water resources around municipal solid waste landfill area in Solan. *MATEC Web. Conf.*, 57, 03011, 2016, EDP Sciences.
22. Zhang, Y., Liu, J., Li, R., Zeng, X., Xue, Y., Liu, J., Nie, Y., Determining the biodegradability of leachate through XAD-8 adsorption. *Procedia Environ. Sci.*, 16, 3–8, 2012.
23. Corsino, S.F., Capodici, M., Di Trapani, D., Torregrossa, M., Viviani, G., Assessment of landfill leachate biodegradability and treatability by means

of allochthonous and autochthonous biomasses. *New Biotechnol.*, 55, 91–97, 2020.
24. Boateng, T.K., Opoku, F., Akoto, O., Heavy metal contamination assessment of groundwater quality: a case study of Oti landfill site, Kumasi. *Appl. Water Sci.*, 9, 2, 33, 2019.
25. Edition, F., Guidelines for drinking-water quality. *WHO Chron.*, 38, 4, 104–8, 2011.
26. Alam, S., Bhardwaj, L.K., Mallick, R., Rai, S., Estimation of heavy metals and fluoride ion in vegetables grown nearby the stretch of River Yamuna, Delhi (NCR), India. *Indian J. Environ. Prot.*, 43, 1, 64–73, 2023.
27. Lewis Sr, R.J., *Hazardous chemicals desk References*, John Wiley & Sons, 2008.
28. Inyang, M., Gao, B., Zimmerman, A., Zhou, Y., Cao, X., Sorption and cosorption of lead and sulfapyridine on carbon nanotube-modified biochars. *Environ. Sci. Pollut. Res.*, 22, 3, 1868–1876, 2015.
29. World Health Organization., *Guidelines for drinking-water quality*, vol. 1, World health organization, 2004.
30. Guerrero, L.A., Maas, G., Hogland, W., Solid waste management challenges for cities in developing countries. *Waste Manage.*, 33, 1, 220–232, 2013.
31. Zurbrügg, C., Drescher, S., Rytz, I., Sinha, A.M.M., Enayetullah, I., Decentralised composting in Bangladesh, a win-win situation for all stakeholders. *Resour. Conserv. Recycl.*, 43, 3, 281–292, 2005.
32. Moghadam, M.A., Mokhtarani, N., Mokhtarani, B., Municipal solid waste management in Rasht City, Iran. *Waste Manage.*, 29, 1, 485–489, 2009.
33. Sharholy, M., Ahmad, K., Mahmood, G., Trivedi, R.C., Municipal solid waste management in Indian cities–A review. *Waste Manage.*, 28, 2, 459–467, 2008.
34. Sharholy, M., Ahmad, K., Vaishya, R.C., Gupta, R.D., Municipal solid waste characteristics and management in Allahabad, India. *Waste Manage.*, 27, 4, 490–496, 2007.
35. Sujauddin, M., Huda, S.M., Hoque, A.R., Household solid waste characteristics and management in Chittagong, Bangladesh. *Waste Manage.*, 28, 9, 1688–1695, 2008.
36. Mrayyan, B. and Hamdi, M.R., Management approaches to integrated solid waste in industrialized zones in Jordan: A case of Zarqa City. *Waste Manage.*, 26, 2, 195–205, 2006.
37. Agamuthu, P., Landfilling in developing countries. *Waste Manage. Res.*, 31, 1, 1–2, 2013.
38. Ethiopia, O., Federal Democratic Republic of Ethiopia, Enhancing Economic Development and Job Creation in Addis Ababa: the Role of the City Administration, 2018.
39. Asase, M., Yanful, E.K., Mensah, M., Stanford, J., Amponsah, S., Comparison of municipal solid waste management systems in Canada and Ghana: A case study of the cities of London, Ontario, and Kumasi, Ghana. *Waste Manage.*, 29, 10, 2779–2786, 2009.

40. Nithya, K., Sathish, A., Kumar, P.S., Ramachandran, T., Fast kinetics and high adsorption capacity of green extract capped superparamagnetic iron oxide nanoparticles for the adsorption of Ni (II) ions. *J. Ind. Eng. Chem.*, 59, 230–241, 2018.
41. Zhang, Y.J., Ou, J.L., Duan, Z.K., Xing, Z.J., Wang, Y., Adsorption of Cr (VI) on bamboo bark-based activated carbon in the absence and presence of humic acid. *Colloids Surf., A*, 481, 108–116, 2015.
42. Ghosh, P., Das, M.T., Thakur, I.S., Mammalian cell line-based bioassays for toxicological evaluation of landfill leachate treated by Pseudomonas sp. ISTDF1. *Environ. Sci. Pollut. Res.*, 21, 13, 8084–8094, 2014.
43. AE, D.L., CA, G.R., García, P., Sandoval, A., Fitorremediación: una alternativa para eliminar la contaminación. *Trop. Subtrop. Agroecosyst.*, 14, 2, 597–612, 2011.
44. Kaveeshwar, A.R., Ponnusamy, S.K., Revellame, E.D., Gang, D.D., Zappi, M.E., Subramaniam, R., Pecan shell based activated carbon for removal of iron (II) from fracking wastewater: adsorption kinetics, isotherm and thermodynamic studies. *Process Saf. Environ. Prot.*, 114, 107–122, 2018.
45. Saravanan, A., Kumar, P.S., Renita, A.A., Hybrid synthesis of novel material through acid modification followed ultrasonication to improve adsorption capacity for zinc removal. *J. Cleaner Prod.*, 172, 92–105, 2018.
46. Pap, S., Knudsen, T.Š., Radonić, J., Maletić, S., Igić, S.M., Sekulić, M.T., Utilization of fruit processing industry waste as green activated carbon for the treatment of heavy metals and chlorophenols contaminated water. *J. Cleaner Prod.*, 162, 958–972, 2017.
47. Jerez Chaverri, J.A., Remoción de metales pesados en lixiviados mediante fitorremediación, 2013.
48. Meng, Q., Chen, H., Lin, J., Lin, Z., Sun, J., Zeolite A synthesized from alkaline assisted pre-activated halloysite for efficient heavy metal removal in polluted river water and industrial wastewater. *J. Environ. Sci.*, 56, 254–262, 2017.
49. Nemati, M., Hosseini, S.M., Shabanian, M., Novel electrodialysis cation exchange membrane prepared by 2-acrylamido-2-methylpropane sulfonic acid; heavy metal ions removal. *J. Hazard. Mater.*, 337, 90–104, 2017.
50. Bhardwaj, L.K., Kumar, D., Kumar, A., Phytoremediation potential of Ocimum sanctum: A sustainable approach for remediation of heavy metals, in: *Phytoremediation Potential of Medicinal and Aromatic Plants*, pp. 46–57, CRC Press, 2023.
51. Rezania, S., Taib, S.M., Din, M.F.M., Dahalan, F.A., Kamyab, H., Comprehensive review on phytotechnology: heavy metals removal by diverse aquatic plants species from wastewater. *J. Hazard. Mater.*, 318, 587–599, 2016.
52. Yu, X.Z., Feng, Y.X., Liang, Y.P., Kinetics of phyto-accumulation of hexavalent and trivalent chromium in rice seedlings. *Int. Biodeterior. Biodegrad.*, 128, 72–77, 2018.

53. Bavandpour, F., Zou, Y., He, Y., Saeed, T., Sun, Y., Sun, G., Removal of dissolved metals in wetland columns filled with shell grits and plant biomass. *Chem. Eng. J.*, 331, 234–241, 2018.
54. Leung, J.Y., Cai, Q., Tam, N.F., Comparing subsurface flow constructed wetlands with mangrove plants and freshwater wetland plants for removing nutrients and toxic pollutants. *Ecol. Eng.*, 95, 129–137, 2016.
55. Papaevangelou, V.A., Gikas, G.D., Tsihrintzis, V.A., Chromium removal from wastewater using HSF and VF pilot-scale constructed wetlands: Overall performance, and fate and distribution of this element within the wetland environment. *Chemosphere*, 168, 716–730, 2017.
56. Costa, A.M., Alfaia, R.G.D.S.M., Campos, J.C., Landfill leachate treatment in Brazil–An overview. *J. Environ. Manage.*, 232, 110–116, 2019.
57. Yabroudi, S.C., Morita, D.M., Alem, P., Landfill leachate treatment over nitritation/denitritation in an activated sludge sequencing batch reactor. *APCBEE Procedia*, 5, 163–168, 2013.

10

Breaking Barriers, Seizing Solutions: Navigating Challenges and Opportunities in Heavy Metal Remediation

Qudrat Ullah[1]*, Laxmi Kant Bhardwaj[2], Waqas Haider[1] and Muhammad Waqar[1]

[1]*Food Technology and Innovation Research Centre of Excellence, School of Agricultural Technology and Food Industry, Walailak University, Nakhon Si Thammarat, Thailand*
[2]*Amity Institute of Environmental Toxicology, Safety and Management, Amity University, Noida, Uttar Pradesh, India*

Abstract

Heavy metal (HM) contamination globally remains a threat to soil, water, and air, as evidenced by the 2023 mining spill of 12 tons of lead (Pb) and cadmium (Cd) in Peru's Cajamarca region. The main focus of discussion in this chapter is the subject of the multiple challenges facing and evolving opportunities for HM remediation, especially in soils contaminated as a result of metals leached from mining, industrial effluents, and agriculturally derived runoff. The key obstacles to treating HMs are the non-biodegradability of HMs and the high cost of treatment. Precise HM estimation can be done by using analytical techniques such as atomic absorption spectroscopy (AAS) and inductively coupled plasma-mass spectrometry (ICP-MS), but they are so far inaccessible. The United Nations Environmental Programme (UNEP), World Health Organization (WHO), and United States Environmental Protection Agency (USEPA) are limited by funding and policy gaps, especially in developing countries. Bioremediation with microbes, phytoremediation with plants (i.e., *Phragmites australis*), and nanoremediation provide promise, along with economic models such as bio-energy crops and circular economy (CE) initiatives. In this chapter, it is argued that the solutions that are advocated should be cost-effective, contextual, and result from interdisciplinary

**Corresponding author*: qudratullahmpur@gmail.com; ORCID: https://orcid.org/0000-0003-3881-0917

Prangya Rath, Laxmi Kant Bhardwaj and Amit Kumar (eds.) Heavy Metal Remediation, (243–262)

collaboration and the evolution of policy to bridge the gaps in the process towards resilient, environment-friendly HM remediation strategies.

Keywords: Bioaccumulation, bioremediation, heavy metals, nanoremediation, phytoremediation, sustainability

10.1 Introduction to Heavy Metal (HM) Contamination and Remediation Challenges

The mining spill occurred in Peru's Cajamarca region in September 2023 when 12 tons of Pb and Cd were spilled in local rivers and were estimated to contaminate 300 hectares of farmland and place 10,000 residents at risk [1, 2]. Besides, this incident transforms into something which highlights the problems of HMs contamination in the soils, water, and air, again targeting several metals like Pb, Cd, chromium (Cr), mercury (Hg), arsenic (As), nickel (Ni) and zinc (Zn) which are contaminating the soil, water and air [2, 3]. Produced from industrial discharges, mining and agricultural runoff, these pollutants are highly detrimental to the environment and health, and urgent, yet highly challenging to remediate [4]. The scope and challenges of HM remediation are investigated in this part, wherein limitations of existing technologies and barriers to viable approaches are critically examined.

Heavy metals (HMs) like Pb, As, and Cd are toxic and bioaccumulative, and disrupt ecosystems and human health [5–7]. HMs are pollutants in soil, water, and air and cause risks to the ecosystem and human health [8–10]. Remediation of HMs faces hurdles like cost, speed, and byproducts. Pb and Cd from mining remain for decades in the soil, where yields of crops are 20 to 40% lower in contaminated areas [11]. Cr (VI) and Hg in water, which are commonly from industrial effluents, are highly mobile, and the global estimate is that 1.2 billion people are drinking HM-contaminated water [12]. Airborne As and Ni from coal combustion are attributed to 3.8 million premature deaths every year, related to respiratory and cardiovascular diseases [13]. A 2023 study in Bangladesh showed that 85% of rice fields near industrial zones had Cd levels exceeding WHO safety thresholds, adding to food chain risks [14]. The statistics of these HMs in various environmental matrices demonstrate the large and pervasive threat of HMs posed by sources such as smelting (Pb and Zn), pesticide use (As), and tannery waste (Cr), among others. The impact of HM pollution on soil, water, and air, highlighting challenges like high costs, slow toxic methods, and byproducts, is supported by visual representations of affected ecosystems (Figure 10.1).

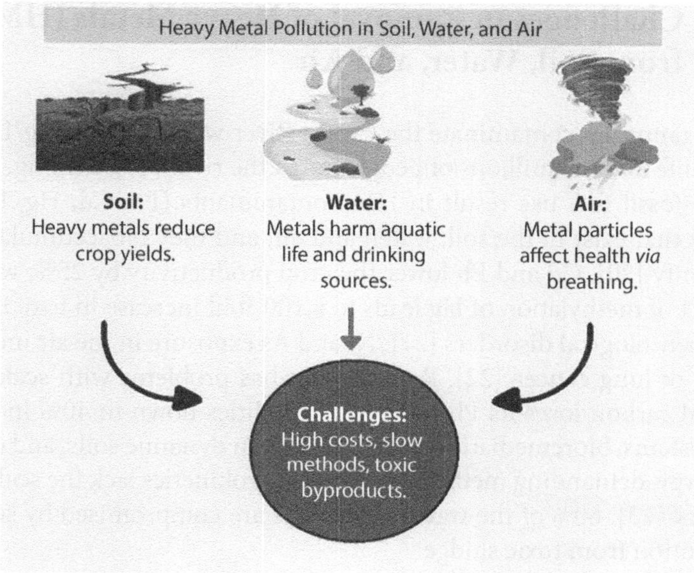

Figure 10.1 Overview of heavy metal (HM) contamination and remediation challenges.

There are several technological, economic, and environmental challenges when it comes to remediating HM contamination. Some conventional methods, such as chemical precipitation and adsorption, are also good at removing Pb from water, but cause toxic sludge and bring secondary pollution [15]. Although phytoremediation using plants such as *Helianthus annuus*, which take up Cd, is sustainable, it is slow, needing 5 to 10 years to generate significant results [16]. Microbes such as *Pseudomonas aeruginosa* used for bioremediation have issues of scalability, which allows field conditions to break the efficacy of the microbes themselves [17]. Finally, electrokinetics were effective for Ni in clay soils, but required high energy inputs, costing \$100 to \$500 per cubic meter. The stark economic barriers to remediation projects in developing nations are reflected in the fact that 70% of remediation projects do not receive funding. The complexity of multi-metal contamination and regulatory inconsistencies additionally limits a successful outcome. As an illustration, an attempt at a 2022 remediation effort in China's Hunan province was unable to tackle mixed Pb-Cr pollution because it is incompatible with such a method [18]. As a result, they require innovative, cost-effective, and context-specific solutions to bridge the gap between scientific potential and real impact.

10.2 Challenges in Removal of Heavy Metals (HMs) from Soil, Water, and Air

Kanpur tanneries contaminate the Ganga River with Cr (15.0 mg/L), killing wildlife and the millions of people down the river [19]. Mining, industry, and fossil fuel use result in HM contaminants (Pb, Cd, Hg, Ni, As, and Zn) that exist in the soil, water, and air, and they bioaccumulate and biomagnify [20]. Cd and Pb lower the crop productivity by 25%, whereas the effect of methylation of Hg leads to a 100-fold increase in toxicity, and causes neurological disorders [21]. Ni and As exposure in the air increases the risk of lung cancer [22]. Remediation has problems with scalability: activated carbon loses its Pb removal capabilities down to 40% in multi-metal systems, bioremediation does not work in dynamic soils, and expensive energy-demanding methods such as electrokinetics lack the scalability necessary [23]. 60% of the water treatments are compromised by secondary pollution from toxic sludge.

Table 10.1 underscores the matrix-specific challenges, revealing why uniform remediation strategies fail. Addressing these requires innovative, context-adapted technologies to overcome persistence, scalability, and pollution risks.

10.3 Techniques for Estimation of Heavy Metal (HM) Contamination

Spillage of 8.0 mg/L Ni from a coal mine in 2021 in South Africa's Mpumalanga province into local waters required urgent monitoring to protect 1.5 million people who live in this area [32, 33]. Estimation of HMs such as Pb, Cd, Cr, Hg, As, Ni, and Zn in soil, water, and air is imperative owing to their hazardous nature. The techniques that are used to measure HMs are usually AAS and ICP-MS, but the limitations of each technique show that they cannot be used for every HM [34, 35].

10.3.1 Analytical Method: Atomic Absorption Spectroscopy (AAS)

Atomic absorption spectroscopy (AAS) is the technique that is used to quantify HM concentrations, whereby metal atoms absorb light in a flame or graphite furnace. The detection limit of monitoring aquatic systems is 0.05 mg/L for Zn in the soil, 0.1 mg/L for Cd, and 0.05 mg/L Pb in the

Table 10.1 Comparison of heavy metal (HM) removal challenges across soil, water, and air.

S. no.	Challenge description	Soil	Water	Air	References
1	High persistence	Cadmium (Cd) binds to clay (50.0 mg/kg)	Mercury (Hg) methylation (2.0 μg/L)	Nickel (Ni) particulates (0.01 μg/m^3)	[24]
2	Bioaccumulation	Lead (Pb) in crops (0.4 mg/kg)	Chromium (Cr) in fish (1.0 mg/kg)	Arsenic (As) in lungs (0.02 μg/g)	[25]
3	Competitive binding	Multi-metal soils reduce adsorption	Lead-chromium (Pb-Cr) competition in water	Mixed particulates	[26]
4	Scalability issues	Bioremediation fails in acidic soils	Membrane fouling in the field	Limited air capture tech	[27]
5	Secondary pollution	Sludge from precipitation	Brine from reverse osmosis (RO) systems	Filter waste	[28]
6	Energy intensity	Electrokinetics (500.0 kWh/m^3)	High-pressure filtration	Scrubber energy use	[29]
7	Cost barriers	$400/ha for bioremediation	$100/m^3 for membrane tech	$500/ton for air filters	[30]
8	Regulatory gaps	Weak enforcement in 40% of regions	Inconsistent water standards	No global air heavy metal (HM) policy	[31]

water [36]. It has been used widely, particularly in resource-limited settings, because it is simple and has a low cost. For example, a study conducted AAS to measure Pb contamination in 1.2 mg/kg of soil, located near smelters, to help improve regulatory action [37]. In contrast with AAS, the analysis of multi-element samples is complicated; a series of measurements must be taken, and matrix interferences lead to lower accuracy in complex samples such as industrial effluents. While adequate to detect high concentrations of Bisphenol A (BPA), this sensitivity is, however, too low to detect trace levels.

10.3.2 Advanced Technique: Inductively Coupled Plasma-Mass Spectrometry (ICP-MS)

Inductively Coupled Plasma-Mass Spectrometry (ICP-MS) ionizes the samples in high-temperature plasma and analyzes the charge ratio for highly sensitive multi-element detection. ICP-MS excels in detecting trace HMs, guiding precise remediation efforts [38]. Its workflow ensures accurate quantification across diverse environmental samples (Figure 10.2). It achieves detection limits as low as 0.001 µg/L for Hg and 0.005 µg/L for As in water, far surpassing AAS. A 2024 study in China used ICP-MS to quantify Cr and Ni at 0.02 $\mu g/m^3$ in air samples, linking emissions to health risks [39]. ICP-MS's precision, with relative standard deviations below 2%, and ability to analyze 20+ elements simultaneously make it ideal for complex matrices. However, its high cost ($100,000–$300,000 per unit) and need for skilled operators limit accessibility in developing nations. Matrix effects, though less severe than in AAS, still require careful calibration. Table 10.2, presented below, compares AAS and ICP-MS across key parameters, illustrating their trade-offs in HM estimation.

Figure 10.2 shows the ICP-MS workflow for HM quantification. The process of ICP-MS, as shown above, is used to measure the concentration of HMs in soil, water, and air, and the diagram thus outlines the step-by-step process, major stages involved, and related challenges.

Table 10.2 highlights ICP-MS's superiority in sensitivity and throughput, critical for multi-metal contamination, while AAS remains practical for single-element, high-concentration scenarios. Accurate HM estimation informs remediation strategies, ensuring targeted interventions. For example, ICP-MS's trace-level detection enabled precise Hg monitoring in South Africa, guiding bioremediation efforts. However, cost and accessibility barriers necessitate strategic deployment of both techniques to balance precision with practicality, ultimately safeguarding ecosystems and public health.

10.4 Techniques for Removal of Heavy Metal (HM) Contamination

Vietnam's textile factory discharged 10.0 mg/L Cr into the Mekong Delta in 2023, necessitating urgent remediation to safeguard the 17.0 million downstream population [45]. Removal of HM from contaminated soil, water, and air is a difficult task around the globe, and conventional as well as sustainable technique offers their distinctions. Coagulation-flocculation

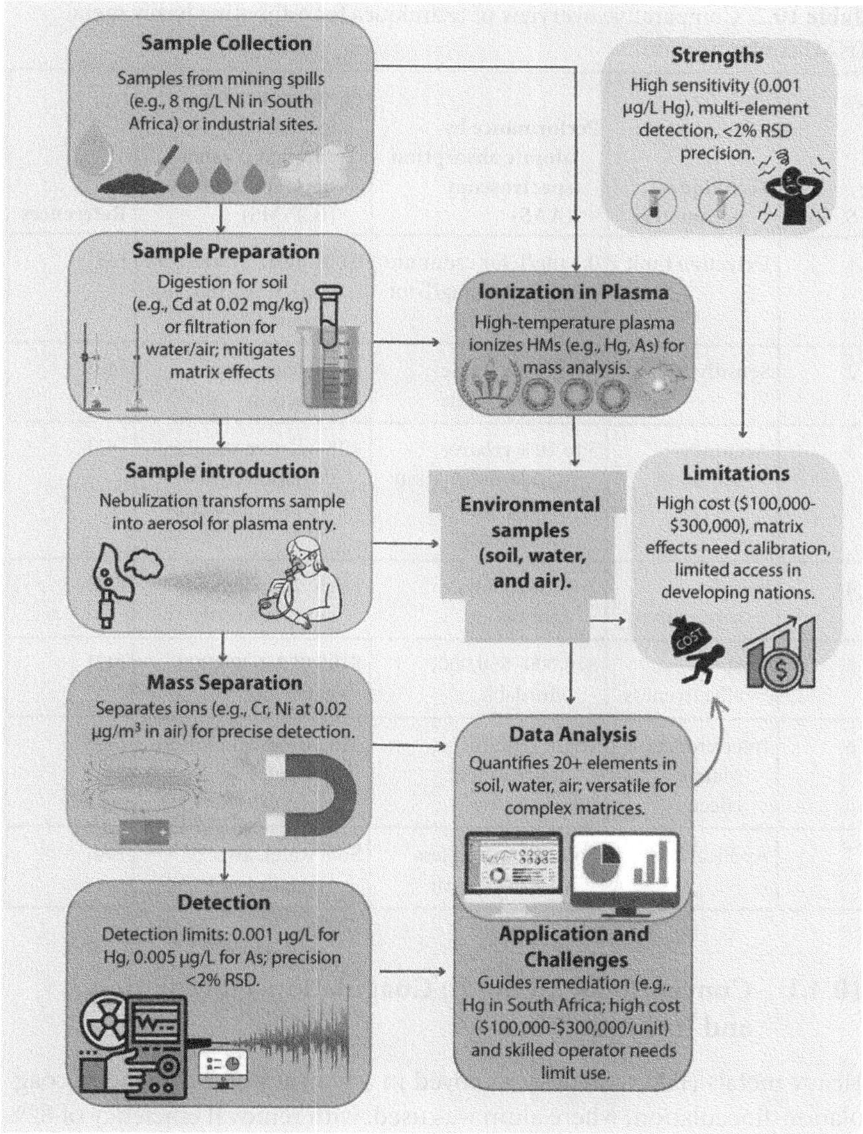

Figure 10.2 Workflow of inductively coupled plasma-mass spectrometry (ICP-MS) for heavy metal (HM) quantification.

is critically evaluated as a traditional method, compared with sustainable approaches (bioremediation, phytoremediation, and nanoremediation) concerning their efficiency, limitations, and potential through real-world data and modeling [46].

Table 10.2 Comparative overview of techniques for estimating heavy metal (HM) contamination.

S. no.	Techniques/ parameters	Performance by atomic absorption spectroscopy (AAS)	Performance by inductively coupled plasma mass spectrometry (ICP-MS)	References
1	Detection limit	0.1 mg/L for cadmium (Cd), 0.05 mg/L for zinc (Zn)	0.001 μg/L for mercury (Hg), 0.005 μg/L for arsenic (As)	[38]
2	Sensitivity	Moderate, single-element focus	High, trace-level detection	[40]
3	Accuracy/ Precision	5 to 10% relative standard deviation (RSD), matrix interference	<2% relative standard deviation (RSD), robust calibration	[41]
4	Sample throughput	Low, sequential analysis	High, 20+ elements simultaneously	[38]
5	Cost-effectiveness	\$10,000–\$50,000, affordable	\$100,000–\$300,000, expensive	[42]
6	Interference/ Matrix effects	High, requires pretreatment	Moderate, calibration mitigates	[43]
7	Applicability	Soil, water, and less effective for air	Soil, water, air; versatile	[44]

10.4.1 Conventional Methods: Coagulation-Flocculation and Its Limitations

Heavy metals (HMs) are also removed in a typical water treatment, coagulation-flocculation, where alum was used, with removal efficiency of 82% Cr (VI) and 60% Pb and Cd, or 90% and 70%, respectively, by the use of chemicals [47]. It yields 0.5 kg/m^3 of sludge to be disposed of (50 \$ 100/ton) and is sensitive to pH (best 6 to 8), which reduces its application in low-resource environments. It is possible to mention sustainable technologies, such as bioremediation using *Pseudomonas putida* (75% of Cd removed in soil), phytoremediation using *Phragmites australis* (67.4% Cd absorption in 6 months), and nanoremediation using iron-nanoparticles (Fe-NPs) (90% As eliminated in 48 hours) [48]. These green procedures minimize secondary pollution and have scalability and cost limitations (\$500/kg of Fe-NPs).

To optimize phytoremediation, the following mass transfer model quantifies HM uptake:

Equation 10.1: Rate of heavy metal (HM) uptake in phytoremediation (Mass Transfer Model)

$$\frac{dc}{dt} = -kA\left(C - C_{eq}\right)$$

Where

- C: HM concentration at time (t) (mg/L)
- C_{eq}: Equilibrium concentration (mg/L, e.g., 0.1 for Cd)
- k: Mass transfer coefficient (0.02 L/m^2/day for *Phragmites*)
- A: Root surface area (0.5 m^2 for mature plants)

A 2024 study in China applied this model to Cd removal in rice paddies, predicting a 60% reduction over 120 days, aligning with observed data [49]. This equation aids in designing phytoremediation systems by forecasting uptake rates and optimizing plant selection. Sustainable methods outperform coagulation-flocculation in ecological safety but require innovation to overcome scalability and cost barriers. Real-world successes, like nanoremediation in Bangladesh's As-contaminated aquifers, underscore their potential [50]. By integrating predictive models like Equation 10.1, remediation can transition from reactive to proactive, balancing efficiency with environmental integrity.

10.5 Bodies Involved in Heavy Metal (HM) Remediation and Their Challenges

In 2024, a Pb smelter leak in Nigeria's Zamfara state contaminated 200 hectares of farmland, exposing 5,000 children to blood Pb levels exceeding 45.0 μg/dL, a threshold linked to cognitive impairment [51]. Such crises bring out the importance of global and national bodies for the resolution of the conduct of HM remediation. Efforts to mitigate HM pollution are driven by UNEP and WHO [52]. USEPA, but have encountered funding, technical, and policy barriers [53].

10.5.1 Global Organizations: United Nations Environmental Programme (UNEP) and World Health Organization (WHO) Initiatives

The Minamata Convention of UNEP, which was presented to 151 countries by 2024, reduced the Hg emission level due to artisanal gold (Au) mining by 15% since 2017 [54]. Its Planet Gold Program has certified $42 million worth of Hg-free Au, although funding shortages crippled action plans. WHO attributes 70% of the deaths that result from exposure to HM every year to Pb and As, amounting to 1.8 million deaths [55, 56]. The 2024 limit of its Pb (5.0 μmol/L) is going unmonitored in 60% of Sub-Saharan Africa. 40% of the projects are funded by the 200-million-dollar budget of UNEP. At the rate of one billion dollars, the Superfund, conducted by USEPA, cleaned up 450 sites, yet 30% of them lack funds. With a budget of 50 million, the Central Pollution Control Board (CPCB) of India is in a position to clean up 10% of the 320 HM sites, and this is dogged by corruption.

Table 10.3, compiled from literature and agency reports, reveals stark funding disparities and systemic barriers. UNEP and WHO struggle globally, while national agencies in developing countries face acute resource shortages. These gaps hinder scalable remediation, particularly in pollution hotspots, necessitating innovative financing and technology transfer to bridge North-South divides.

10.6 Opportunities for Overcoming Remediation Challenges

In 2024, a Cd spill in China's Hunan province contaminated 500 hectares of farmland, threatening rice yields for 1 million residents [64]. Such crises highlight the urgent need for innovative, sustainable solutions to HM remediation. Traditional methods, hindered by high costs and secondary pollution, are giving way to green technologies and economic models that promise scalability and environmental safety. Cutting-edge remediation technologies and economic incentives like bio-energy crops and circular economy frameworks, offering hope for transformative HM management.

10.6.1 Innovations in Sustainable Remediation Technologies

Real breakthroughs in green chemistry and biotechnology are going to revolutionize HM remediation. Despite high immobilization (95%) efficiency performed in 48h, the addition of these engineered biochars with

Table 10.3 Funding and resource challenges for remediation bodies.

S. no.	Remediation body	Country/ region	Annual budget	Key challenges	References
1	United Nations Environmental Programme (UNEP)	Global	USD 200 M	Donor dependency, project underfunding	[57]
2	World Health Organization (WHO)		USD 150 M	Limited monitoring capacity	[31]
3	United States Environmental Protection Agency (USEPA)	United States of America (USA)	USD 1.2 B	Budget cuts, site prioritization	[58]
4	Central Pollution Control Board (CPCB)	India	USD 50 M	Weak enforcement, corruption	[59]
5	Pakistan Environmental Protection Agency (PEPA)	Pakistan	USD 10 M	Lack of advanced technology	[60]
6	National Environment Management Authority (NEMA)	Nigeria	USD 5 M	Insufficient funding, corruption	[31]
7	South African National Biodiversity Institute (SANBI)	South Africa	USD 8 M	Technical expertise shortage	[61]
8	Consejo Nacional del Ambiente (CONAM)	Peru	USD 12 M	Policy fragmentation	[62]
9	Ministry of Environment, Forest and Climate Change (MoEFCC)	Bangladesh	USD 6 M	Climate vulnerabilities	[63]

functional groups costs 10 \$/kg, much lower than chemical washing (50 \$/kg). Iron oxide nanoparticles (Fe_3O_4-NPs), used for nanoremediation, showed promising results in a pilot test in 2023; they removed 98.4% Cd^{2+} from the groundwater in 6 hours, but scalability is still limited by production costs [65]. A cost-effective and environmentally sustainable approach utilizing bioelectrochemical systems, including microbial fuel cells, treatment of Cr (VI) contaminated wastewater accomplished a 90% reduction for \$5/m^3, while generating 0.2 kWh/m^3 electricity [66]. These technologies enhance environmental safety by minimizing toxic byproducts and adapt to diverse conditions, such as acidic soils or high-salinity water. However, field-scale deployment requires cost reductions and regulatory support to ensure resilience.

10.6.2 Economic Incentives: Bio-Energy Crops and Circular Economy (CE)

Economic models leveraging bio-energy crops and CE principles are transforming HM remediation into a profitable endeavor. *Miscanthus x giganteus*, grown on Cd-contaminated lands, yielded 15 t/ha of biomass for bioethanol, extracting 0.5 kg/ha Cd annually, with farmers earning \$200/ha in a 2023 Brazilian trial [67, 68]. In Thailand, a 2024 CE initiative recycled HM-laden sludge into construction materials, reducing landfill waste by 22% and generating \$1.5 million in revenue [69]. These models repurpose contaminated biomass and waste, aligning with sustainability goals. However, upfront costs and market access for bio-energy products remain barriers, particularly in developing nations. Emerging technologies and economic models are reshaping HM remediation. Nanoremediation achieves 98.4% Cd removal in 6 hours, costing \$500/kg for NPs. Engineered biochar immobilizes 95% Pb in soils at \$10/kg, offering a low-cost alternative. Microbial consortia reduce 80% of Hg in water, while bioreactors achieve 90% Cr (VI) removal, generating 0.2 kWh/m^3 electricity [70, 71]. Economically, *Miscanthus x giganteus* yields 15 t/ha biomass on Cd-contaminated lands, extracting 0.5 kg/ha Cd and earning \$200/ha [72]. CE models in Thailand recycled HM-laden sludge, reducing landfill waste by 22% and generating \$1.5 million [73, 74]. Subsidies (\$100/ha) and carbon credits (\$50/t carbon dioxide (CO_2)) further incentivize adoption.

10.7 Recommendations

Bioremediation, phytoremediation, and nano remediation applications should be used as an interdisciplinary thread to promote HM regulation with the aim of offering ecologically friendly solutions at an economically viable scale by using cost-effective alternatives (engineered biochar, 10 USD/kg; microbial consortia). Regulations should be simplified, and sustainable steps encouraged with the use of subsidies ($100/ha) and carbon credits ($50/t CO_2) by policymakers. Investments in readily available analysis tools, such as AAS and local ICP-MS systems, can improve HM surveillance in developing countries. Mechanisms like recycling of HM-laden sludge (into construction material) as CE models should be implemented on larger scales, and the remediation plans must be resilient and context-specific, such that ecosystems and the health of people are not compromised.

10.8 Conclusion: Bridging Challenges with Sustainable Innovations

Remediation of HM is experiencing problems such as expensive coagulation-flocculation, generation of harmful sludge, and difficulty in dealing with complex pollution. The lab-based solution is affected by an inconsistent regulatory environment and funding gaps, particularly among developing countries. Eco-friendly solutions of sustainable alternatives can be applied in bioremediation by using microbes, phytoremediation with plants, and nanoremediation capable of localizing the site of contaminant immobilization. Miscanthus and other bioenergy crops have a CE and turn waste into resources, recovering the ecosystem and making money. Previous measures must require cheap engineered biochar, microbial consortium, and inter-sector cooperation, so that the technology can be appropriately transferred. Policy incentives and simplification of regulations should be a priority to fuel proactive remediation, moving toward a healthier and cleaner planet on a larger scale.

List of Abbreviations

AAS	Atomic absorption spectroscopy
As	Arsenic
Au	Gold
BPA	Bisphenol A
Cd	Cadmium
CE	Circular economy
CONAM	Consejo Nacional del Ambiente
CO_2	Carbon dioxide
CPCB	Central Pollution Control Board
Cr	Chromium
Fe_3O_4-NPs	Iron oxide nanoparticles
Fe-NPs	Iron nanoparticles
Hg	Mercury
HM	Heavy metal
ICPMS	Inductively coupled plasma mass spectrometry
MoEFCC	Ministry of Environment, Forest and Climate Change
NEMA	National Environment Management Authority
Ni	Nickel
Pb	Lead
PEPA	Pakistan Environmental Protection Agency
RO	Reverse osmosis
RSD	Relative standard deviation
SANBI	South African National Biodiversity Institute
UNEP	United Nations Environmental Programme
USA	United States of America
USEPA	United States Environmental Protection Agency
WHO	World Health Organization
Zn	Zinc

AI disclosure statement: The author confirms that no artificial intelligence (AI)-generated content or AI-assisted writing tools were used in the preparation of this chapter (including summaries and the abstract). All text, analysis, and conclusions are the result of the author's original work.

References

1. Cáceda, C.J., Maraza, G., de Lourdes Fora, G., Farfan, D.G., Obando, E., Chiampo, F., Carpio, M., Evaluation of Potentially Toxic Elements in Soil and Potential Ecological Risks Generated by Environmental Liabilities in Tacna, Peru, bioRxiv, 2024-09, 2024.
2. Ccanccapa-Cartagena, A., Chavez-Gonzales, F.D., Paredes, B., Vera, C., Gutierrez, G., Valencia, R., Alcázar, A.L.P., Zyaykina, N.N., Filley, T.R., Jafvert, C.T., Seasonal differences in trace metal concentrations in the major rivers of the hyper-arid southwestern Andes basins of Peru. *J. Environ. Manage.*, 344, 118493, 2023.
3. La-Cruz-Flores, H.V., Zárate-Quiñones, R.H., Carbajal-Morán, H., Osorio-Berrocal, E., Mazzarino, M.J., Use of the Organic Fraction of Urban Solid Waste to Recover Degraded Areas in Chilca, Peru. *Pol. J. Environ. Stud.*, 34, 1, 2025.
4. Liu, X., Sathishkumar, K., Zhang, H., Saxena, K.K., Zhang, F., Naraginti, S., Rajendiran, R., Rajasekar, A., Guo, X., Frontiers in environmental cleanup: Recent advances in remediation of emerging pollutants from soil and water. *J. Hazard. Mater. Adv.*, 16, 100461, 2024.
5. Alam, S., Bhardwaj, L.K., Mallick, R., Rai, S., Estimation of heavy metals and fluoride ion in vegetables grown nearby the stretch of River Yamuna, Delhi (NCR), India. *Indian J. Environ. Prot.*, 43, 1, 64–73, 2023.
6. Bhardwaj, L.K., Sharma, S., Jindal, T., Estimation of physico-chemical and heavy metals in the lakes of Grovnes & Broknes Peninsula, Larsemann Hill, East Antarctica. *Chem. Afr.*, 6, 5, 2677–2694, 2023.
7. Gupta, U., Shukla, S., Bhardwaj, L.K., Rath, P., Upadhyay, D., Sharma, B., Rupam, Asthana, H., Jindal, T., Singh, H., Unraveling urban plant strategies for heavy metal uptake and detoxification, in: *Urban Forests, Climate Change and Environmental Pollution: Physio-Biochemical and Molecular Perspectives to Enhance Urban Resilience*, pp. 93–119, Springer Nature Switzerland, Cham, 2024.
8. Monib, A.W., Niazi, P., Azizi, A., Sediqi, S., Baseer, A.Q., Heavy metal contamination in urban soils: health impacts on humans and plants: a review. *Eur. J. Theor. Appl. Sci.*, 2, 1, 546–565, 2024.
9. Bhardwaj, L.K., Rath, P., Choudhury, M., Metzdorff, A., Jodder, P.K., Electronic wastes and contamination of water, in: *Electronic Waste*, pp. 48–57, CRC Press, London, UK, 2025.
10. Bhardwaj, L.K., Rath, P., Jain, H., Choudhury, M., Exploring the effects of e-waste on soil, water quality and human health. *Discov. Civ. Eng.*, 2, 1, 12, 2025.
11. Yu, Q., Gao, B., Zhang, X., Agricultural wastes improve soil quality and enhance the phytoremediation efficiency of economic crops for heavy metal-contaminated soils in mining areas. *Environ. Geochem. Health*, 47, 3, 65, 2025.

12. Wang, L., Wang, Y., Ma, F., Tankpa, V., Bai, S., Guo, X., Wang, X., Mechanisms and reutilization of modified biochar used for removal of heavy metals from wastewater: a review. *Sci. Total Environ.*, 668, 1298–1309, 2019.
13. Goswami, R. and Neog, N., Heavy metal pollution in the environment: impact on air quality and human health implications, in: *Heavy Metal Toxicity: Environmental Concerns, Remediation and Opportunities*, pp. 75–103, Springer Nature Singapore, Singapore, 2023.
14. Akter, M., Kabir, M.H., Alam, M.A., Al Mashuk, H., Rahman, M.M., Alam, M.S., Brodie, G., Islam, S.M., Gaihre, Y.K., Rahman, G.K.M.M., Geospatial Visualization and Ecological Risk Assessment of Heavy Metals in Rice Soil of a Newly Developed Industrial Zone in Bangladesh. *Sustainability*, 15, 9, 7208, 2023.
15. Sarker, A., Al Masud, M.A., Deepo, D.M., Das, K., Nandi, R., Ansary, M.W.R., Islam, A.R.M.T., Islam, T., Biological and green remediation of heavy metal contaminated water and soils: A state-of-the-art review. *Chemosphere*, 332, 138861, 2023.
16. Tan, H.W., Pang, Y.L., Lim, S., Chong, W.C., A state-of-the-art of phytoremediation approach for sustainable management of heavy metals recovery. *Environ. Technol. Innovation*, 30, 103043, 2023.
17. Khoshru, B., Fallah Nosratabad, A., Mahjenabadi, V.A.J., Knežević, M., Hinojosa, A.C., Fadiji, A.E., Enagbonma, B.J., Qaderi, S., Patel, M., Baktash, E.M., Dawood, M.F.A.M., Multidimensional role of Pseudomonas: from biofertilizers to bioremediation and soil ecology to sustainable agriculture. *J. Plant Nutr.*, 48, 6, 1016–1042, 2025.
18. Hao, H., Liu, F., Huang, Z., Lv, Y., Chen, W., Ge, D., Spatial Distribution, Contamination Feature, and Health-Risk Assessment of Heavy Metals and Metalloids in Farmland Soil in the Hunan Province, China. *J. Environ. Eng.*, 148, 9, 04022053, 2022.
19. Samanta, S., Metal and pesticide pollution scenario in Ganga River system. *Aquat. Ecosyst. Health Manage.*, 16, 4, 454–464, 2013.
20. Jagaba, A.H., Lawal, I.M., Birniwa, A.H., Affam, A.C., Usman, A.K., Soja, U.B., Saleh, D., Hussaini, A., Noor, A., Yaro, N.S.A., Sources of water contamination by heavy metals, in: *Membrane Technologies for Heavy Metal Removal from Water*, pp. 3–27, CRC Press, Basel, Switzerland, 2024.
21. Mondal, S., Mukherjee, S.K., Hossain, S.T., Exploration of plant growth promoting rhizobacteria (PGPRs) for heavy metal bioremediation and environmental sustainability: Recent advances and future prospects, in: *Modern Approaches in Waste Bioremediation: Environmental Microbiology*, Springer Nature, Cham, Switzerland, pp. 29–55, 2023.
22. Lee, N.W., Wang, H.Y., Du, C.L., Yuan, T.H., Chen, C.Y., Yu, C.J., Chan, C.C., Air-polluted environmental heavy metal exposure increase lung cancer incidence and mortality: A population-based longitudinal cohort study. *Sci. Total Environ.*, 810, 152186, 2022.

23. Pandey, B.K., Shukla, C., Sillanpää, M., Shukla, S.K., A systematic review on application of electrokinetics in stabilization and remediation of problematic soils. *Innovative Infrastruct. Solutions*, 8, 9, 226, 2023.
24. Wang, P., Shen, X., Qiu, S., Zhang, L., Ma, Y., Liang, J., Clay-based materials for heavy metals adsorption: Mechanisms, advancements, and future prospects in environmental remediation. *Crystals*, 14, 12, 1046, 2024.
25. Ali, Z., Yousafzai, A.M., Sher, N., Muhammad, I., Nayab, G.E., Aqeel, S.A.M., Shah, S.T., Aschner, M., Khan, I., Khan, H., Toxicity and bioaccumulation of manganese and chromium in different organs of common carp (Cyprinus carpio) fish. *Toxicol. Rep.*, 8, 343–348, 2021.
26. Xu, W., Zheng, J., Cui, M., Lai, H., Enzyme-Induced Carbonate Precipitation for the Stabilization of Heavy Metal-Contaminated Landfill Soils: A Sustainable Approach to Resource Recovery and Environmental Remediation. *Sustainability*, 17, 10, 4630, 2025.
27. Kuppan, N., Padman, M., Mahadeva, M., Srinivasan, S., Devarajan, R., A comprehensive review of sustainable bioremediation techniques: Eco friendly solutions for waste and pollution management. *Waste Manage. Bull.*, 2, 3, 154–171, 2024.
28. Bratovcic, A., Buksek, H., Helix-Nielsen, C., Petrinic, I., Concentrating hexavalent chromium electroplating wastewater for recovery and reuse by forward osmosis using underground brine as draw solution. *Chem. Eng. J.*, 431, 133918, 2022.
29. Kurt, S., Energetic Optimisation of Wastewater Treatment Plant and Evaluation of Green House Gases, Doctoral dissertation, Politecnico di Torino, 2024.
30. Sheoran, K., Siwal, S.S., Kapoor, D., Singh, N., Saini, A.K., Alsanie, W.F., Thakur, V.K., Air pollutants removal using biofiltration technique: a challenge at the frontiers of sustainable environment. *ACS Eng. Au.*, 2, 5, 378–396, 2022.
31. Ogbeide, O. and Henry, B., Addressing heavy metal pollution in Nigeria: Evaluating policies, assessing impacts, and enhancing remediation strategies. *J. Appl. Sci. Environ. Manage.*, 28, 4, 1007–1051, 2024.
32. Mudau, K.L., An assessment of water quality and occurrence of antibiotic-resistant bacteria in Naauwpoortspruit River, Mpumalanga province, South Africa, University of South Africa, South Africa, 2021.
33. Mamoepa, B.M., An ecofeminist analysis of environmental justice in South Africa: Mpumalanga Highveld and Cape Town region, Doctoral dissertation, Stellenbosch University, Stellenbosch, 2024.
34. Evans, E.H., Pisonero, J., Smith, C.M., Taylor, R.N., Atomic spectrometry update: review of advances in atomic spectrometry and related techniques. *J. Anal. At. Spectrom.*, 36, 5, 868–891, 2021.
35. Yang, H.S., LaFrance, D.R., Hao, Y., Elemental testing using inductively coupled plasma mass spectrometry in clinical laboratories: an ACLPS critical review. *Am. J. Clin. Pathol.*, 156, 2, 167–175, 2021.

36. Rai, S.K., Xalxo, R., Patle, T.K., Verma, A., Chauhan, R., Mahish, P.K., Analyzing contamination of heavy metals-AAS and fluorescence spectroscopy, in: *Heavy Metals in the Environment: Management Strategies for Global Pollution*, pp. 167–204, American Chemical Society, Amsterdam, Netherlands, 2023.
37. Adnan, M., Xiao, B., Ali, M.U., Xiao, P., Zhao, P., Wang, H., Bibi, S., Heavy metals pollution from smelting activities: A threat to soil and groundwater. *Ecotoxicol. Environ. Saf.*, 274, 116189, 2024.
38. Lata, S. and Ansari, N.G., Analytical Techniques for Heavy Metal Analysis, in: *Heavy Metal Contamination in the Environment*, pp. 180–202, CRC Press, Weinheim, Germany, 2024.
39. Abidin, A.U., Maziya, F.B., Susetyo, S.H., Yoneda, M., Matsui, Y., Heavy metal air pollution in an Indonesian landfill site: Characterization, sources, and health risk assessment for informal workers. *Environ. Adv.*, 15, 100512, 2024.
40. Inobeme, A., Mathew, J.T., Jatto, E., Inobeme, J., Adetunji, C.O., Muniratu, M., Onyeachu, B.I., Adekoya, M.A., Ajai, A.I., Mann, A., Olori, E., Recent advances in instrumental techniques for heavy metal quantification. *Environ. Monit. Assess.*, 195, 4, 452, 2023.
41. Bacon, J.R., Butler, O.T., Cairns, W.R., Cavoura, O., Cook, J.M., Davidson, C.M., Mertz-Kraus, R., Atomic spectrometry update–a review of advances in environmental analysis. *J. Anal. At. Spectrom.*, 36, 1, 10–55, 2021.
42. Madheswaran, D.K., Saravanan, J., Jegadheeswaran, S., Subramanian, V., Harnessing biogas in India: a sustainable pathway for sustainable energy development. *Green Low-Carbon Econ.*, 2024.
43. Urge, A.Y., Salting-Out Assisted Liquid-Liquid Extraction for Simultaneous Determination of Steroid Hormones and Thyroxine by Using LC-MS/MS: Application in Fish Plasma, 2022.
44. Wang, Z., Luo, P., Zha, X., Xu, C., Kang, S., Zhou, M., Nover, D., Wang, Y., Overview assessment of risk evaluation and treatment technologies for heavy metal pollution of water and soil. *J. Cleaner Prod.*, 379, 134043, 2022.
45. Tong, N.X. and Thao, N.T.T., Assessment of water quality, heavy metal pollution and human health risks in the Canal system of Ho Chi Minh City, Vietnam. *Environ. Res. Commun.*, 6, 7, 075007, 2024.
46. Olawade, D.B., Wada, O.Z., Egbewole, B.I., Fapohunda, O., Ige, A.O., Usman, S.O., Ajisafe, O., Metal and metal oxide nanomaterials for heavy metal remediation: novel approaches for selective, regenerative, and scalable water treatment. *Front. Nanotechnol.*, 6, 1466721, 2024.
47. Saleh, T.A., Mustaqeem, M., Khaled, M., Water treatment technologies in removing heavy metal ions from wastewater: A review. *Environ. Nanotechnol. Monit. Manage.*, 17, 100617, 2022.
48. Verma, A., Roy, A., Bharadvaja, N., Remediation of heavy metals using nanophytoremediation, in: *Advanced Oxidation Processes for Effluent Treatment Plants*, pp. 273–296, Elsevier, Boca Raton, FL, USA, 2021.

49. Gai, X., Xing, W., Cheng, W., Xiao, J., Chen, G., Regulatory mechanism of bamboo biochar dosage on cadmium accumulation in Salix psammophila: insights from rhizosphere microbial communities, assembly processes, and interactions. *Carbon Res.*, 3, 1, 72, 2024.
50. Bhattacharya, S., Talukdar, A., Sengupta, S., Das, T., Dey, A., Gupta, K., Dutta, N., Arsenic contaminated water remediation: A state-of-the-art review in synchrony with sustainable development goals. *Groundwater Sustainable Dev.*, 23, 101000, 2023.
51. Anka, S.A., Lead poisoning in Zamfara State Nigeria: effects on environmental health, in: *Poisoning-Prevention, Diagnosis, Treatment and Poison Repurposing*, IntechOpen, Berlin, Germany, 2024.
52. Awewomom, J., Dzeble, F., Takyi, Y.D., Ashie, W.B., Ettey, E.N.Y.O., Afua, P.E., Sackey, L.N., Opoku, F., Akoto, O., Addressing global environmental pollution using environmental control techniques: a focus on environmental policy and preventive environmental management. *Discov. Environ.*, 2, 1, 8, 2024.
53. Ulibarri, N., Figueroa, O.P., Grant, A., Barriers and opportunities to incorporating environmental justice in the National Environmental Policy act. *Environ. Impact Assess. Rev.*, 97, 106880, 2022.
54. Donkor, A.K., Ghoveisi, H., Bonzongo, J.C.J., Use of metallic mercury in artisanal gold mining by amalgamation: a review of temporal and spatial trends and environmental pollution. *Minerals*, 14, 6, 555, 2024.
55. Alliance, A.G. and Gold, W., Mercury Poisoning, 2021, https://amazonaid.org/threats-to-the-amazon/mercury-poisoning/.
56. Chowdhury, F.N. and Rahman, M.M., Source and distribution of heavy metal and their effects on human health, in: *Heavy Metal Toxicity: Human Health Impact and Mitigation Strategies*, pp. 45–98, Springer Nature Switzerland, Cham, 2024.
57. O'Brien, R.M., Phelan, T.J., Smith, N.M., Smits, K.M., Remediation in developing countries: A review of previously implemented projects and analysis of stakeholder participation efforts. *Crit. Rev. Environ. Sci. Technol.*, 51, 12, 1259–1280, 2021.
58. Kueper, B.H., Stroo, H.F., Vogel, C.M., Ward, C.H. (Eds.), *Chlorinated solvent source zone remediation*, p. 713, Springer, New York, 2014.
59. Turaga, R.M.R. and Sugathan, A., Environmental regulations in India, in: *Oxford Research Encyclopedia of Environmental Science*, Elsevier, Oxford, UK, 2020.
60. Iqbal, J., Raza, M.H., Imran, M., Ullah, I., Hayat, M., Tayyab, M., ul Islam, M., A Critical Analysis of Implementation of Economic & Industrial Development Strategies in Pakistan. *J. Public Policy*, 2, 2, 2023.
61. Ouma, K.O., Shane, A., Syampungani, S., Aquatic ecological risk of heavy-metal pollution associated with degraded mining landscapes of the Southern Africa River Basins: a review. *Minerals*, 12, 2, 225, 2022.

62. Cruzado-Tafur, E., Assessment of accumulation of selected metals by native plants growing in polluted peruvian post-mining areas, Doctoral dissertation, Université de Pau et des Pays de l'Adour; Pontificia universidad católica del Perú, 2021.
63. Das, L.C., Mohiul Islam, A.S.M., Aktar, S., Sultana, F., Climate change challenges in Bangladesh. *Int. J. Big Data Min. Global Warming*, 6, 01, 2430001, 2024.
64. Rouhani, A., Gutiérrez, M., Newton, R.A., Al Souki, K.S., An overview of potentially toxic element pollution in soil around lead–zinc mining areas. *Environ. Rev.*, 33, 1–16, 2024.
65. Saxena, J., Singh, A., Jyoti, A. (Eds.), *Nanobiotechnology: principles and applications*, Bentham Science Publishers, Singapore, 2023.
66. Aarthy, M., Rajesh, T., Thirunavoukkarasu, M., Critical review on microbial fuel cells for concomitant reduction of hexavalent chromium and bioelectricity generation. *J. Chem. Technol. Biotechnol.*, 95, 5, 1298–1307, 2020.
67. Abdalla, M., Hastings, A., Campbell, G., Chen, H., Smith, P., Assessing the sustainability of Miscanthus and willow as global bioenergy crops: Current and future climate conditions (Part 1). *Agronomy*, 14, 12, 3020, 2024.
68. Anne, O., Blom, A., Faidušienė, R., Karčauskienė, D., Lundström, J., Mockevičienė, I., Rogala, A., Rybarczyk, P., Svensson, S., Schmieder, F., Öhman, D., Guidelines: Phytoremediation technologies and methods for polluted soils treatment, 2023.
69. Franklin-Wallis, O., *Wasteland: The Dirty Truth about what We Throw Away, where it Goes, and why it Matters*, Simon and Schuster, San Diego, CA, USA, 2023.
70. Zand, A.D., Tabrizi, A.M., Heir, A.V., Incorporation of biochar and nanomaterials to assist remediation of heavy metals in soil using plant species. *Environ. Technol. Innovation*, 20, 101134, 2020.
71. Dubey, P., Farooqui, A., Patel, A., Srivastava, P.K., Microbial innovations in chromium remediation: mechanistic insights and diverse applications. *World J. Microbiol. Biotechnol.*, 40, 5, 151, 2024.
72. Peddyreddy, S.R., Industrial agriculture: complication to solution [online], 2023.
73. Farooq, A., Haputta, P., Gheewala, S.H., Economic feasibility assessment of waste to energy technologies for the development of a sustainable municipal solid waste management system in Thailand. *Renewable Energy*, 233, 121155, 2024.
74. Samitthiwetcharong, S., Chavalparit, O., Suwanteep, K., Murayama, T., Kullavanijaya, P., Enhancing circular plastic waste management: Reducing GHG emissions and increasing economic value in Rayong province, Thailand. *Heliyon*, 10, 18, 2024.

11

Charting the Horizon: Future Innovations and Global Strategies for Heavy Metal Remediation

Qudrat Ullah[1]*, Laxmi Kant Bhardwaj[2], Waqas Haider[1] and Tassaneem Yuttasa-ard[1]

[1]Food Technology and Innovation Research Centre of Excellence, School of Agricultural Technology and Food Industry, Walailak University, Nakhon Si Thammarat, Thailand
[2]Amity Institute of Environmental Toxicology, Safety and Management, Amity University, Noida, Uttar Pradesh, India

Abstract
Heavy metal (HM) contamination, menacing ecosystem health and human health, is collectively threatened by industrialization, urbanization, and climate change at a global level, with an increase of about 30% by 2030. This chapter finally discusses future perspectives and innovative strategies for HM remediation with special reference to nanotechnology (NT), metagenomics, and the development of global policy frameworks. Nanoremediation by nano zeolites and iron oxide nanoparticles (Fe_3O_4-NPs) immobilizes up to 65% of lead (Pb) in soils, and hybrid electrocoagulation–microbial systems remove up to 85% of cadmium (Cd) from wastewater, surpassing classical processes. Instead, metagenomics serves as a tool for microbial bioremediation, that is, a reduction of HM concentrations in water and in the air. Farming yields drop by 20%, and there is a 40% loss of biodiversity in aquatic systems; human health risks for neurological and carcinogenic risks increase by 25%, especially for vulnerable populations. Technology transfer and knowledge sharing are being driven under the United Nations Sustainable Development Goals (UNSDGs) and also through initiatives like the European Union's (EU's) Green Deal, but in regions such as the Middle East, technology transfer is hindered due to gaps in enforcement. Finally, among the strategic

**Corresponding author*: qudratullahmpur@gmail.com; ORCID: https://orcid.org/0000-0003-3881-0917

Prangya Rath, Laxmi Kant Bhardwaj and Amit Kumar (eds.) Heavy Metal Remediation, (263–284)

recommendations they came up with, stricter regulations should be issued on phytoextraction use, more funding for low-cost phytoremediation, and being able to get artificial intelligence (AI) driven monitoring to ensure sustainable remediation, and keeping the ecosystem and the people's health in the future.

Keywords: Biochar, electrocoagulation, heavy metals, metagenomics, nanoremediation, phytoremediation

11.1 Introduction

Climate change, industrialization, and urbanization have led to the threat of HM contamination to ecosystems, human lives, and sustainable development, with foreseen rates of increase of 30% by 2030 [1]. Uncontrolled mining, electronic waste (e-waste) leaching, and coal burning are increasing the pollutants of Pb, Cd, and arsenic (As) in the soil, water, and air [2]. In 2030, the value of Pb (150.0 mg/kg) in Southeast Asian soils is estimated to increase by 35%, whereas urban river basins may be confronted with 20% higher Cd loads (>5.0 to 25.0 μg/L) [3]. Urban Pb could also rise by 15% in the air to aggravate respiratory problems. The soil erosion and flooding group enhances HM mobility by 40% fewer fish deaths in the polluted rivers. The neurological disorders and cancer risks increase within the vulnerable groups due to long-term exposure to As and Cd by 25% [4]. Conventional remediation is no longer highly efficient, although it has been replaced by promoters such as nanoremediation, phytoremediation, and green chemistry, which are more productive. When such NPs were added to the soil, the NPs of zero-valent iron (Fe) used in India lowered the amount of As by 60% in 2024. In 2023, *Brassica juncea* in the soil in China removed 300.0 mg/kg Pb [5]. The insertion of biochar reduced water Cd by 45% in 2025 [6]. Byproducts-recycled-to-remediation agents, strategies such as the circular economy (CE), maximize sustainability, and overcome the pressing need to find methods of scale at which to stave off the aftereffects of HM pollution. To mitigate the long-term ecological and health risks of HM contamination, these innovations need the support of global policies that are in line with the UNSDGs. Acting preemptively is how to ensure natural resources and human well-being for future generations. This is an increase in the growing problem of HM pollution from industrial and urban sources (Figure 11.1). Simple, effective solutions like nature-based methods and innovative technologies to tackle the problem are also displayed.

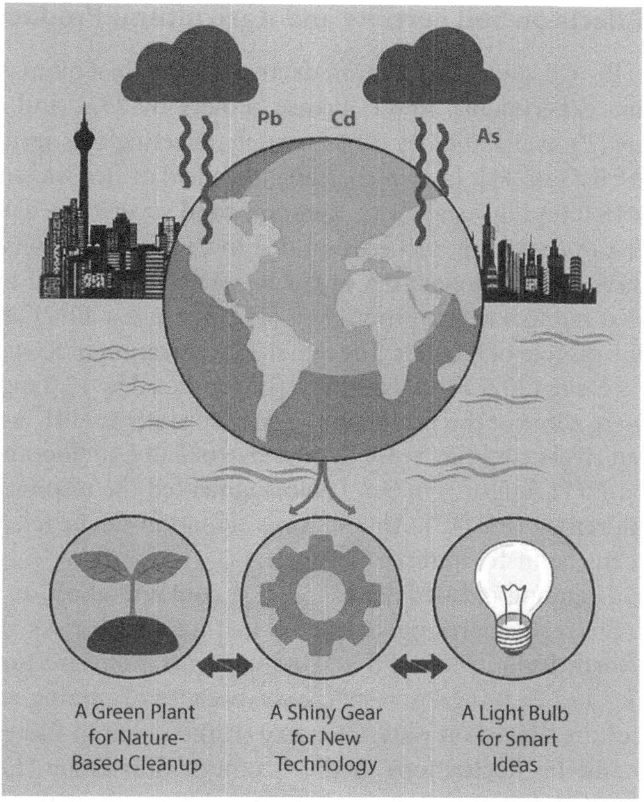

Figure 11.1 Future heavy metal (HM) challenges and solutions with simple ways.

11.2 Future Environmental Impacts of Heavy Metal (HM) Contamination

Heavy metals (HMs) contamination is rising all over the world, and their environmental effects are expected to get worse: soil fertility, aquatic ecosystem, and air quality are on the verge of being threatened [7–9]. These impacts are driven by industrial expansion, urbanization, and climate change and are of urgent concern, including the cascading effects of HMs on critical environmental systems, trends projected under future HMs scenarios, and potential mitigation strategies for safeguarding our planet's future.

11.2.1 Effects on Soil Fertility and Agricultural Productivity

The metal (Pb, Cd, and As) pollution decreases microbial biomass by 30% (2024 China experiment), lowers urease activity by 25%, and decreases phosphorus (P) availability by 20%, severely affecting the agriculture of Asia and Africa [10, 11]. In Punjab, India, the yield of rice was suppressed by 15% by 18.0 mg/kg As at a rate that puts food security in danger [12]. On the African continent, soil erosion due to climate and monsoon conditions causes an additional 10% of Cd to leach into the soil by 2025, a situation that reduces maize production [13]. There is a 40% Cd bioavailability in the sample of biochar. The lethal concentration of 8.0 μg/L Cd in the Mekong Delta (2023) and the health risks induced by 12.0 mg/kg of Pb in tilapia were some of the results in the aquatic systems [14]. As leaching in European rivers could increase by 15% by 2025 due to flooding [3]. Air transported Pb (1.5 $\mu g/m^3$) in San Paolo augmented the respiration problems of children by 18% [15]. Air and water pollution can be fixed with the help of Green chemistry and scrubbers.

The trends among industrial areas in HM contamination, as shown in Table 11.1, are really alarming. Scenarios that have high risks are in East Asia and North India, where Pb and As concentrations are projected at 150.0 mg/kg and 25.0 mg/kg, respectively, because of mining and runoff from agriculture. Based on water mercury (Hg) content in Eastern Russia (6.0 μg/L) and Pb fluctuations in the Southern Brazil's air (1.8 $\mu g/m^3$), severe dangers to aquatic life and urban atmospheric air quality problems were determined, respectively. And these projections highlight the necessity for targeted interventions in high–risk zones.

11.3 Future Effects on Human Health

Industrialization, urbanization, and environmental degradation have increased the presence of HMs such as Pb, Cd, and As to levels that seriously affect human health [25]. According to its projections, the burden of HM-related disease will increase sharply by 2030, especially in industrial hot spots and in developing countries. The neurological and carcinogenic health risk trends associated with long-term HM exposure, as well as how vulnerable populations disproportionately suffer the consequences of exposure, and critical trends and potential interventions are discussed (Figure 11.2).

Table 11.1 Projected heavy metal (HM) concentration trends in industrial regions.

S. no.	Region/ location	Key heavy metals (HMs)	Projected concentration in 2030	Environmental risk category	References
1	East Asia	Lead (Pb) and cadmium (Cd)	150.0 mg/kg of lead (Pb), 22.0 mg/kg of cadmium (Cd)	High	[16]
2	Northern India	Arsenic (As) and cadmium (Cd)	25.0 mg/kg of arsenic (As), 18.0 mg/kg of cadmium (Cd)		[17]
3	Central Africa	Lead (Pb) and mercury (Hg)	90.0 mg/kg of lead (Pb), 5.0 mg/kg of mercury (Hg)	Moderate	[18]
4	Western Europe	Arsenic (As) and chromium (Cr)	10.0 µg/L of arsenic (As), 15.0 µg/L of chromium (Cr)		[19]
5	Southern Brazil	Lead (Pb) and nickel (Ni)	1.8 $\mu g/m^3$ of lead (Pb), 0.9 $\mu g/m^3$ of nickel (Ni)	High	[20]
6	Midwest United States of America (USA)	Cadmium (Cd) and zinc (Zn)	12.0 mg/kg of cadmium (Cd), 200.0 mg/kg of zinc (Zn)	Moderate	[21]
7	Southeast Australia	Arsenic (As) and copper (Cu)	20.0 mg/kg of arsenic (As), 150.0 mg/kg of copper (Cu)		[22]
8	Eastern Russia	Mercury (Hg) and lead (Pb)	6.0 µg/L of mercury (Hg), 80.0 mg/kg of lead (Pb)	High	[23, 24]

11.3.1 Long-Term Exposure Risks: Neurological and Carcinogenic Effects

Long-term exposure to HMs such as Pb and Cd is the most deleterious in their neurological and carcinogenic impacts involving bioaccumulation and cytotoxicity [26]. Pb blocks the N-methyl-D-aspartate (NMDA) receptors, which play an essential role in learning and remembering, and China research 2024 demonstrates that at a blood level of 10.0 g/dL (twice the World Health Organization (WHO) recommendation of 5.0 g/dL), the children have a 20%

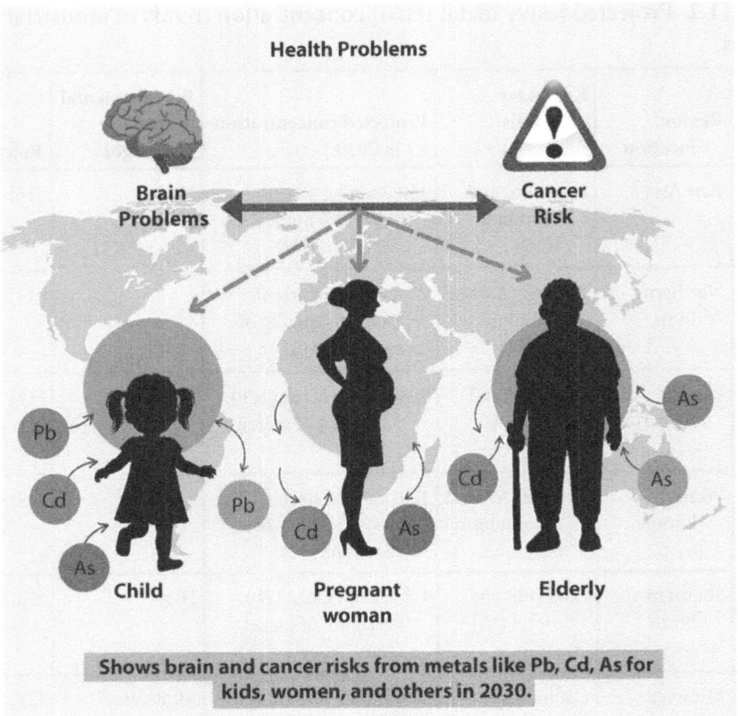

Figure 11.2 Health risks from heavy metals (HMs): brain and cancer risks associated with metals like lead (Pb), cadmium (Cd), and arsenic (As) for children, women, and other populations.

lower cognitive score [27]. Cd causes oxidative deoxyribonucleic acid (DNA) damage and reactive oxygen species (ROS), thereby increasing cancer risk. A 2024 study in Bangladesh associated Cd concentrations in drinking water (50.0 μg/L, over 16 times the WHO guideline of ≤3.0 μg/L) with a 30% increase in lung and bladder cancer risk over 10 years [28]. An endocrine disruption by Cd also increases the likelihood of type 2 diabetes by 25%, even though the urinary level of this chemical is at 2.0 μg/L. WHO forecasts that the number of the disease burden linked to HM will grow by 15% in the next 2035 years, and of course, more enforceable policies and medical attention are essential to decrease prolonged toxicity [29].

11.3.2 Impacts on Vulnerable Populations

The vulnerable groups are exposed to a greater risk of exposure to HM through air, water, soil, and food, such as infants, pregnant women, and the elderly. Developmental newborn toxicity was increased by 40% in South Africa, with pregnant women living near gold (Au) mines having 8.0 μg/L of blood Hg [30]. Among the natives of Canada, as in fish (15.0 mg/kg), raised skin cancer amongst the native individuals by 22% [31]. The urban children exposed to Pb in dust (1.2 g/m^3) and water (20.0 g/L) were found to have 35% more behavioral disorders in Nigeria [32]. In Poland, neurodegenerative diseases in the elderly increased by a factor of 28% due to atmospheric Cd (0.8 g/m^3) [33]. Mapping of the exposure using AI and equal access to resources are crucial to reducing these health hazards.

Table 11.2 reveals stark trends in HM-related health risks. Urban children in India and Nigeria face cognitive and behavioral challenges due to

Table 11.2 Projected health impact trends due to heavy metal (HM) exposure.

S. no.	Region/ population group	Dominant heavy metals (HMs)	Projected health outcomes (2030+)	High-risk factors	References
1	Urban children, India	Lead (Pb) and mercury (Hg)	Cognitive decline, behavioral disorders	Contaminated water, roadside pollution	[34]
2	Pregnant women, South Africa	Mercury (Hg) and arsenic (As)	Developmental toxicity, low birth weight	Mining runoff, contaminated fish	[35]
3	Indigenous adults, Canada	Arsenic (As) and cadmium (Cd)	Skin cancer, kidney dysfunction	Contaminated fish, soil exposure	[36]
4	Low-income children, Nigeria	Lead (Pb) and chromium (Cr)	Behavioral disorders, anemia	Roadside dust, polluted water	[37]
5	Elderly, Poland	Cadmium (Cd) and nickel (Ni)	Neurodegenerative diseases, respiratory issues	Atmospheric deposition, urban air	[38]
6	Rural farmers, Vietnam	Arsenic (As) and lead (Pb)	Liver cancer, neurological impairment	Irrigation water, pesticide runoff	[39]
7	Urban adults, Brazil	Lead (Pb) and mercury (Hg)	Cardiovascular diseases, hypertension	Industrial emissions, urban water	[40]
8	Children, Bangladesh	Arsenic (As) and cadmium (Cd)	Growth retardation, lung cancer risk	Drinking water, rice consumption	[41]

Pb exposure *via* water and dust, while pregnant women in South Africa are at risk of developmental toxicity from Hg. Indigenous groups in Canada and rural farmers in Vietnam show elevated cancer risks from bioaccumulation in food chains. Urban adults in Brazil and elderly populations in Poland have, for the first time, escalated the load of neurodegenerative and cardiovascular diseases caused by atmospheric HMs. This highlights the importance of establishing strong policy frameworks (e.g., stricter water quality standards, community health monitoring programs) to minimize the environmental inequities that result in an unequal burden of HM exposure.

11.4 Emerging Technologies for Heavy Metal (HM) Removal

The remediation of HM contamination has provided a persistent challenge that has highlighted the shortcomings of the conventional treatment techniques, including chemical precipitation and phytoremediation, which face the limitations of scale-up, cost, and secondary pollution [42, 43]. However, metals such as Pb and Cd are leaching from soil, water, and air as human activity continues with industrial development and urbanization [44]. This contention is supported by emerging technologies, especially in the field of NT and metagenomics, which would present transformative ways of tackling these issues.

11.4.1 Nanoremediation: Immobilizing Heavy Metals (HMs) in Soil

Nanoremediation is perfect to immobilize HMs in soil, minimizing their leaching to the surroundings. Fe_3O_4-NPs, nano-zeolite, and nano-biochar are found to provide good surface area and reactivity. A 2024 Brazilian study demonstrated that Fe_3O_4-NPs (10 to 20 nm, 120.0 m^2/g) removed 65% of Pb from groundwater within a pH range of 5.5 to 6.5, protecting microbial activity [45]. In 2023, a Vietnamese study reported that nano-zeolite (150.0 m^2/g) immobilized 70% of Cd in acidic soils (pH 4.0 to 5.5), enhancing soil remediation [46]. Reduction of Cr bioavailability by 55%, maintaining the fertility of the soil, was also achieved due to nano-biochar application.

11.4.2 Metagenomics and Hybrid Methods for Water and Air

Metagenomics and hybrid technologies change HM elimination in water and air. A 2021 Nigerian study employed metagenomic sequencing to identify bacteria that were able to remove Cd contents in wastewater down to 1.5 mg/L, achieving 85% efficiency [47]. Clustered Regularly Interspaced Short Palindromic Repeats (CRISPR) increases HM tolerance in microbes to increase bioremediation. Metagenomic biosensors identify airborne Pb to enable to implementation of specific interventions [48]. An Indian test in 2020 used a combination of electrocoagulation and microbial degradation, with a Cr initial concentration of 15.0 mg/L and a final concentration of 0.8 mg/L, in 48 hours, consuming 2.5 kWh/m^3, which is 30% less energy than that of conventional electrocoagulation, requiring 4.0 kWh/m^3 with its 60% efficiency [49].

Equation 11.1: Efficiency model for hybrid heavy metal (HM) removal techniques [50, 51]
The efficiency of hybrid HM removal systems can be quantified using the following equation:

$$E\,(\%) = \left(Ci - \frac{Cf}{Ci} \right) \times 100$$

where

- E is the removal efficiency (%)
- Ci is the initial HM concentration (mg/L)
- Cf is the final HM concentration (mg/L)

This model assumes steady-state conditions and uniform mixing in the treatment system. Chaudharı, Chotalıya [52] Applied this equation to evaluate Cr removal. With ($Ci = 15.0$) mg/L and ($Cf = 0.8$) mg/L, the efficiency was calculated as:

$$E = \left(\frac{15 - 0.8}{15} \right) \times 100 = 94.7\%$$

Electrocoagulation followed by microbial action exhibits this high efficiency, which outperforms traditional methods. Increasingly, such models are used for the optimization of operational parameters for real-time control of wastewater treatment to assure compliance with environmental standards.

11.5 Policies and Regulations for Sustainable Remediation

Heavy metal (HM) contamination remains a silent but pervasive threat to degrading ecosystems and human health the world over, particularly due to poorly enforced or rather inadequate policies [53–55]. Soil, water, and air polluted with metals like Pb and Cd are the legacy of industrial discharges and mining, and the consequences are catastrophic. So, for example, in India, industrial effluents are only treated and disinfected before being discharged into the environment, eventually resulting in increased Pb levels of up to 80.0 μg/L (eight times the WHO limit value) in groundwater, resulting in the rise in chronic kidney disease in the affected communities by 15% [56, 57]. It then looks at incentives and international agreements shaping HM remediation technology as viewed through the lens of global HM policies and how these align with sustainable development goals (SDGs) in the context of harmonized governance [58].

11.5.1 Global Standards: Alignment with United Nations Sustainable Development Goals (UNSDGs)

Heavy metal (HM) cleanup is directed by the UNSDGs, which are poorly enforced. SDG 3 reports that 60% of the low-income countries do not have Pb limits, worsening the neurological disorders of children by 12%. SDG 6 is faced with a pollution of Cd in the Vietnamese rivers (10.0 μg/L compared to 3.0 μg/L recommended by WHO) [59, 60]. Middle Eastern unsustainable management of wastes results in the emission of 200 tons of HMs per year (SDG 12), whereas the percentage of flooding increases soil As by 20% in the Association of Southeast Asian Nations (ASEAN) region (SDG 13) [61]. Among the African laws, 30% of the arable land is degraded (SDG 15) [62]. The European Union (EU) green deal of 2023 (500 M), Chinese cleansing measure of 2024 (40%), and phytoremediation in ASEAN all have an effect of 30%, 40%, and 15% reductions in Pb, Cd, and Hg, respectively, in soil [63]. The United States of America (USA) electrocoagulation of 300 M improves the water quality by 25%, whereas the fund emptiness in the Middle East increases the atmospheric Pb by 10% [64]. Scalable solutions that are driven through capacity building include the 2024 United Nations Environment Programme (UNEP) Technology Transfer Fund and the 2025 EU-ASEAN Pact [65].

As shown in Table 11.3, there are different regional approaches to the remediation of HM. The EU and China do best with robust funding and technology adoption to reduce soil HMs by large amounts. ASEAN and the USA focus on scalable remediation solutions like phytoremediation and electrocoagulation. More so, the Middle East and South Africa have enforcement gaps and a lack of funding, which has a limiting effect. As most of these frameworks highlight the avenue for cross-sectoral

Table 11.3 Policy frameworks supporting heavy metal (HM) remediation.

S. no.	Country/region	Key regulation/policy	Focus area	Technology/ approach promoted	References
1	European Union (EU)	2024 Green Deal Expansion	Soil	Bioremediation	[66]
2	China	2024 Soil Pollution Action Plan		Nano-biochar	[67]
3	Association of Southeast Asian Nations (ASEAN) (Thailand)	2023 Regional Framework for Mining		Phytoremediation	[68]
4	United States of America (USA)	2024 Clean Water Act Amendment	Water	Electrocoagulation	[69]
5	Middle East, Kingdom of Saudi Arabia (KSA)	2023 Vision 2030 Environmental Plan	Air	Emission controls	[70]
6	India	2024 National Water Policy Update	Water	Adsorption systems	[71]
7	South Africa	2022 Mining Rehabilitation Act	Soil	Bioenergy crops	[72]
8	Vietnam	2022 Industrial Effluent Standards	Water	Microbial remediation	[73]
9	Brazil	2025 Amazon Reforestation Initiative	Soil	Phytoremediation	[74]
10	Australia	2023 Clean Air Strategy	Air	Scrubber technology	[75]

collaboration and financial incentives, they also need to address the disparities in enforcement in order to avoid inequitable progress.

11.6 The Role of Global Collaboration in Future Remediation

Industrialization ushers in a global crisis: HM contamination, which calls for unified action. Soil Pb levels in some regions exceed 200.0 mg/kg, and UNEP in 2024 finds that the stakes are high in pollution in Asia's rivers and Africa's mining scars that threaten food security and health [76]. International collaboration provides a lifeline, through shared knowledge and technology transfer, which accelerates remediation.

11.6.1 Sharing Best Practices across Regions

Global organizations such as UNEP, United Nations Industrial Development Organization (UNIDO), and Organization for Economic Co-operation and Development (OECD) promote the HM remediation process based on knowledge transfer. UNEP 2024 initiative in Thailand transferred the practice of phytoremediation to Vietnam, making soil Cd levels decrease by 35% (12.0 to 7.8 mg/kg) through the use of Salix species at 20% reduced costs [77]. South Africa-Kenya partnership initiative of the African Ministerial Conference on the Environment (AMCEN) embraced biochar in 2023 and reduced the soil Pb by 40% (150.0 to 90.0 mg/kg) and enhanced the quality of water by 15% [78]. Poland was aided by the EU 2024 Horizon Program, in which Bacillus reduced Cr in wastewater to half, from 10.0 to 5.0 mg/L [79]. Technology transfers are affected by issues such as the lack of funds and training. A 60% Hg decrease in Zimbabwe, funded by the Global Environment Facility (GEF), deployed nano-zeolite, and a 45% decrease in soil As using a non-governmental organization (NGO) funded *Brassica juncea* in a 2022 project in Bangladesh [80, 81]. Although in Ethiopia, electrocoagulation did not work, the World Bank will provide USD 200 M in 2024 to fund SDG-compatible technology transfers, but with a focus on multi-stakeholder work to ensure sustainable, up-scalable solutions [82].

11.7 Recommendations

To combat the growing HM pollution, the world needs to consider the incorporation of nanoremediation (e.g., Fe_3O_4-NPs, 500/kg), phytoremediation (e.g., *Brassica juncea*), and metagenomics as efficient means to rid of Pb and Cd by up to 65 to 85%. The policymakers need to strengthen the industrial standards on emissions, provide funding ($100/ha) to support the low cost of biochar ($10/kg) and phytoextraction. With geospatial mapping by AI, it is possible to increase the accuracy by 25%, focusing on zones at high risk, such as East Asia. In line with UNSDGs, a collaborative effort between countries, as well as the transformation of technology to countries such as the Middle East, will prompt sustainable and equitable clean-up without jeopardizing the human population and other ecological systems.

11.8 Conclusion: A Roadmap for Sustainable Heavy Metal (HM) Remediation

Climate change, industrial growth, and urbanization increase HM pollution in the environment, and by 2030, pollution is expected to increase by 30%, which will reduce soil fertility by 20%, aquatic biodiversity by 40%, and neurological and cancer risks among susceptible groups by 25%. The solutions are emerging technologies to remove Pb up to 65% (nano-zeolites) and Cd up to 85% (metagenomics), yet international joint efforts and SDG-based policies play a critical role. Tightening regulations can reduce HM emissions by 50% by treating them in electrocoagulation and biochar, and the international funding and training can help treat 70% of polluted areas with phytoextraction. Using AI monitoring, there is a 25% increase in the efficiency of responses, and sustainable remediation is achieved by multi-stakeholder efforts that can be sustained in future generations.

List of Abbreviations

AI	Artificial intelligence
AMCEN	African Ministerial Conference on the Environment
As	Arsenic
ASEAN	Association of Southeast Asian Nations
Au	Gold
Cd	Cadmium
CE	Circular economy
Cr	Chromium
CRISPR	Clustered Regularly Interspaced Short Palindromic Repeats

Cu	Copper
DNA	Deoxyribonucleic acid
EU	European Union
E-waste	Electronic waste
Fe	Iron
Fe_3O_4-NPs	Iron oxide nanoparticles
GEF	Global Environment Facility
Hg	Mercury
HM	Heavy metal
KSA	Kingdom of Saudi Arabia
NGO	Non-governmental organization
Ni	Nickel
NMDA	N-methyl-D-aspartate
NPs	Nanoparticles
NT	Nanotechnology
OECD	Organization for Economic Co-operation and Development
P	Phosphorus
Pb	Lead
ROS	Reactive oxygen species
SDGs	Sustainable Development Goals
UNEP	United Nations Environmental Programme
UNIDO	United Nations Industrial Development Organization
UNSDGs	United Nations Sustainable Development Goals
USA	United States of America
WHO	World Health Organization
Zn	Zinc

AI disclosure statement: The author confirms that no artificial intelligence (AI)-generated content or AI-assisted writing tools were used in the preparation of this chapter (including summaries and the abstract). All text, analysis, and conclusions are the result of the author's original work.

References

1. Adnan, M., Xiao, B., Xiao, P., Zhao, P., Bibi, S., Heavy metal, waste, COVID-19, and rapid industrialization in this modern era—Fit for sustainable future. *Sustainability*, 14, 8, 4746, 2022.
2. Abubakar, A., Zangina, A.S., Maigari, A.I., Badamasi, M.M., Ishak, M.Y., Abdullahi, A.S., Haruna, J.A., Pollution of heavy metal threat posed by

e-waste burning and its assessment of human health risk. *Environ. Sci. Pollut. Res.*, 29, 40, 61065–61079, 2022.

3. Alfee, S.L. and Bloor, M.C., A global review of river sediment contamination and remobilization through climate change-induced flooding. *Sustainable Environ.*, 11, 1, 2440957, 2025.
4. Agnihotri, S.K. and Kesari, K.K., Mechanistic effect of heavy metals in neurological disorder and brain cancer, in: *Networking of Mutagens in Environmental Toxicology*, pp. 25–47, Springer International Publishing, Cham, 2019.
5. Sehrish, A.K., Ahmad, S., Alomrani, S.O., Ahmad, A., Al-Ghanim, K.A., Alshehri, M.A., Tauqeer, A., Ali, S., Sarker, P.K., Nutrient strengthening and lead alleviation in Brassica Napus L. by foliar ZnO and TiO2-NPs modulating antioxidant system, improving photosynthetic efficiency and reducing lead uptake. *Sci. Rep.*, 14, 1, 19437, 2024.
6. Shoudho, K.N., Khan, T.H., Ara, U.R., Khan, M.R., Shawon, Z.B.Z., Hoque, M.E., Biochar in global carbon cycle: Towards sustainable development goals. *Curr. Res. Green Sustainable Chem.*, 8, 100409, 2024.
7. Ijaz, S., Iqbal, J., Abbasi, B.A., Tufail, A., Ullah, Z., Yaseen, T., Ali, I., Uddin, S., Iqbal, R., Heavy metal–polluted arable land and its consequences: A global scenario, in: *Biochar-assisted remediation of contaminated soils under changing climate*, pp. 71–99, IntechOpen, Tehran, Iran, 2024.
8. Alam, S., Bhardwaj, L.K., Mallick, R., Rai, S., Estimation of heavy metals and fluoride ion in vegetables grown nearby the stretch of River Yamuna, Delhi (NCR), India. *Indian J. Environ. Prot.*, 43, 1, 64–73, 2023.
9. Gupta, U., Shukla, S., Bhardwaj, L.K., Rath, P., Upadhyay, D., Sharma, B., Rupam, A.H., Jindal, T., Singh, H., Unraveling urban plant strategies for heavy metal uptake and detoxification, in: *Urban Forests, Climate Change and Environmental Pollution: Physio-Biochemical and Molecular Perspectives to Enhance Urban Resilience*, pp. 93–119, Springer Nature Switzerland, Cham, 2024.
10. Farhad, M., Noor, M., Yasin, M.Z., Nizamani, M.H., Turan, V., Iqbal, M., Interactive suitability of rice stubble biochar and arbuscular mycorrhizal fungi for improving wastewater-polluted soil health and reducing heavy metals in peas. *Sustainability*, 16, 2, 634, 2024.
11. da Silva Martins, T., Garcia, K.G.V., da Silva, Y.J.A.B., da Silva, M.G., Serpa, S.S.E., Bezerra, R.A., Filho, C.D.T., Cavalcante, R.M., Boechat, C.L., de Araujo Pereira, A.P., Escobar, M.E.O., Contamination risk by heavy metals and enzymatic stoichiometry in agricultural soils under intense use of pesticides. *Environ. Monit. Assess.*, 196, 9, 805, 2024.
12. Kaur, M., Malik, D.P., Malhi, G.S., Sardana, V., Bolan, N.S., Lal, R., Siddique, K.H., Rice residue management in the Indo-Gangetic Plains for climate and food security. A review. *Agron. Sustainable Dev.*, 42, 5, 92, 2022.

13. Igwe, E., Assessing sub-Saharan Africa's GHG emissions from croplands: environmental impacts and sustainable mitigation strategies. *Environ. Monit. Assess.*, 197, 2, 208, 2025.
14. Sallam, G.R., Aly, H.A., Lotfy, A.M., Abdel-Rahim, M.M., Fayed, W.M., Teiba, I.I., Mzengereza, K., Tembo, M., Singini, W., Habib, Y.J., Shehata, A.I., Natural zeolite for heavy metal, ammonia removal, and physiological responses in European sea bass (Dicentrarchus labrax) juveniles tanks with different densities. *PLoS One*, 19, 4, e0297844, 2024.
15. Sannoh, F., Khwaja, H.A., Fatmi, Z., Rizvi, N.A., Turabi, A., Hussain, M.M., Siddique, A., Carpenter, D.O., Gaseous pollutants linked to pulmonary diseases: East meets West. *Air Qual. Atmos. Health*, 17, 12, 2879–2897, 2024.
16. Satarug, S., Environmental Cadmium Exposure Limits and Thresholds for Adverse Effects on Kidneys. *J. Environ. Expose. Assess.*, 2025.
17. Rani, J. and Paul, B., Ecological risk assessment of heavy metals in soils of lignite mining area of Kutch district of Gujarat, India. *Environ. Monit. Assess.*, 196, 8, 766, 2024.
18. Adebote, M.O., Aggrey-Fynn, J., Mensah, P.K., Assessing mercury and lead pollution in the Ankobra estuary due to artisanal mining activities: Implications for water quality and aquatic life. *PLoS One*, 20, 6, e0325909, 2025.
19. Dippong, T., Mihali, C., Avram, A., Evaluating groundwater metal and arsenic content in Piatra, North-West of Romania. *Water*, 16, 4, 539, 2024.
20. de Almeida Piai, K., Nogueira, T., Kaneshiro Olympio, K.P., Nardocci, A.C., Assessment of human health risks associated with airborne arsenic, nickel and lead exposure in particulate matter from vehicular sources in Sao Paulo city. *Int. J. Environ. Health Res.*, 34, 4, 1926–1943, 2024.
21. Livingston, V., Trace Metal Partitioning in Environmental Compartments in Nahant Marsh, Davenport Iowa, Doctoral dissertation, Western Illinois University, 2024.
22. Poggere, G., Gasparin, A., Barbosa, J.Z., Melo, G.W., Corrêa, R.S., Motta, A.C.V., Soil contamination by copper: Sources, ecological risks, and mitigation strategies in Brazil. *J. Trace Elem. Miner.*, 4, 100059, 2023.
23. Yakovlev, E., Zykova, E., Zykov, S., Druzhinina, A., Ivanchenko, N., Evaluation of heavy metal pollution of snow and groundwater on the territory of suburban community garden plots of the Arkhangelsk Agglomeration (Northwest Russia). *Pollution*, 8, 4, 1448–1473, 2022.
24. Batyrova, G., Kononets, V., Umarova, G., Taskozhina, G., Umarov, Y., Issanguzhina, Z., Kudabayeva, K., Batyrov, R., Trace Metal and Metalloid Profiles in Hair Samples from Children in the Oil-Producing Region of Kazakhstan. *Toxics*, 13, 7, 522, 2025.
25. Chowdhury, F.N. and Rahman, M.M., Source and distribution of heavy metal and their effects on human health, in: *Heavy metal toxicity: human health impact and mitigation strategies*, pp. 45–98, Springer Nature Switzerland, Cham, 2024.

26. Mitra, S., Chakraborty, A.J., Tareq, A.M., Emran, T.B., Nainu, F., Khusro, A., Idris, A.M., Khandaker, M.U., Osman, H., Alhumaydhi, F.A., Simal-Gandara, J., Impact of heavy metals on the environment and human health: Novel therapeutic insights to counter the toxicity. *J. King Saud Univ. Sci.*, 34, 3, 101865, 2022.
27. Wang, Y., Huang, Y., Ma, A., You, J., Miao, J., Li, J., Natural antioxidants: an effective strategy for the treatment of Alzheimer's disease at the early stage. *J. Agric. Food Chem.*, 72, 21, 11854–11870, 2024.
28. Parvin, F., Haque, M.M., Tareq, S.M., Recent status of water quality in Bangladesh: a systematic review, meta-analysis and health risk assessment. *Environ. Challenges*, 6, 100416, 2022.
29. Ding, Y., Deng, A., Qi, T.F., Yu, H., Wu, L.P., Zhang, H., The burden and trend prediction of ischemic heart disease associated with lead exposure: Insights from the Global Burden of Disease study 2021. *Environ. Health*, 24, 1, 1–19, 2025.
30. Sylvester, A.A., Health Risk Evaluation and the Investigation of the Potential Human Foetal Subacute Developmental Toxicities upon Exposures to Mercury-Contaminated Food Crops from Artisanal and Small-Scale Gold Mining Communities–A Case Study of Obuasi, Ghana, Doctoral dissertation, 2023.
31. Andrade-Rivas, F., Afshari, R., Yassi, A., Mardani, A., Taft, S., Guttmann, M., Rao, A.S., Thomas, S., Takaro, T., Spiegel, J.M., Industrialization and food safety for the Tsleil-Waututh Nation: An analysis of chemical levels in shellfish in Burrard Inlet. *Environ. Res.*, 206, 112575, 2022.
32. Ewona, I.O., Ekah, B.J., Akwagiobe, J.U., Udo, S.O., Rabiu, B., Evaluation of Hazard Quotient (Hq) of Toxic and Heavy Metal Concentration in Air Sediments Across Selected Cities in Nigeria, 2024.
33. Doroszkiewicz, J., Farhan, J.A., Mroczko, J., Winkel, I., Perkowski, M., Mroczko, B., Common and trace metals in Alzheimer's and Parkinson's diseases. *Int. J. Mol. Sci.*, 24, 21, 15721, 2023.
34. Hanif, S., Ali, S., Chaudhry, A.H., Sial, N., Marium, A., Mehmood, T., A review on soil metal contamination and its environmental implications. *Nature Environ. Pollut. Technol.*, 24, 1, 1–14, 2025.
35. Adebowale, A.Q., Heavy metal and organic compound bioaccumulation in bronze whaler sharks (Carcharhinus brachyurus) along the coastline of South Africa, 2024.
36. Chan, H.M., Singh, K., Batal, M., Marushka, L., Tikhonov, C., Sadik, T., Schwartz, H., Ing, A., Fediuk, K., Levels of metals and persistent organic pollutants in traditional foods consumed by First Nations living on-reserve in Canada. *Can. J. Public Health*, 112, Suppl 1, 81–96, 2021.
37. Kolawole, T.O., Oyelami, C.A., Olajide-Kayode, J.O., Jimoh, M.T., Fomba, K.W., Anifowose, A.J., Akinde, S.B., Contamination and risk surveillance of potentially toxic elements in different land-use urban soils of Osogbo, Southwestern Nigeria. *Environ. Geochem. Health*, 45, 7, 4603–4629, 2023.

38. Majewski, G., Klik, B., Rogula-Kozłowska, W., Rogula-Kopiec, P., Rybak, J., Radziemska, M., Liniauskienė, E., Assessment of heavy metal inhalation risks in urban environments in Poland: A case study. *J. Ecol. Eng.*, 24, 11, 2023.
39. Olson, K.R. and Cihacek, L., Agent blue spraying in the Mekong Delta during the Vietnam war: Fate of the arsenic based herbicide weapon used to destroy rice crop and mangrove forests. *Open J. Soil Sci.*, 12, 7, 253–294, 2022.
40. Crespo-Lopez, M.E., Lopes-Araújo, A., Basta, P.C., Soares-Silva, I., de Souza, C.B., Leal-Nazaré, C.G., Santos-Sacramento, L., Barthelemy, J.L., Arrifano, G.P., Augusto-Oliveira, M., Environmental pollution challenges public health surveillance: the case of mercury exposure and intoxication in Brazil. *Lancet Reg. Health Am.*, 39, 2024.
41. Khan, M.I., Ahmad, M.F., Ahmad, I., Ashfaq, F., Wahab, S., Alsayegh, A.A., Kumar, S., Hakeem, K.R., Arsenic exposure through dietary intake and associated health hazards in the Middle East. *Nutrients*, 14, 10, 2136, 2022.
42. Bhardwaj, L.K., Kumar, D., Kumar, A., Phytoremediation potential of Ocimum sanctum: A sustainable approach for remediation of heavy metals, in: *Phytoremediation potential of medicinal and aromatic plants,* pp. 46–57, Scrivener Publishing, Salem, Massachusetts, USA, 2023.
43. Bahrdwaj, L.K., Upadhyay, D., Jain, H., Rath, P., Kumar, D., Bajpai, S., Recent trends of phytoremediation for the wastewater treatment and resource recovery, in: *Biotechnologies for wastewater treatment and resource recovery,* pp. 135–146, Scrivener Publishing, Salem, Massachusetts, USA, 2025.
44. Khatun, J., Intekhab, A., Dhak, D., Effect of uncontrolled fertilization and heavy metal toxicity associated with arsenic (As), lead (Pb) and cadmium (Cd), and possible remediation. *Toxicology*, 477, 153274, 2022.
45. Bilal, M., Iqbal, H.M., Gupta, R.K., Nguyen, T.A., Yasin, G. (Eds.), *Nanotechnology to monitor, remedy, and prevent pollution,* John Wiley & Sons, Hoboken, New Jersey, USA, 2024.
46. Xu, H., Ou, Z., Li, W., Hu, T., Zhang, Y., Xu, H., Wang, J., Li, Y., Cadmium (II) adsorption by recyclable Zeolite-Loaded Hydrogel: Extension to the removal of Cadmium (II) from contaminated soil. *Chem. Eng. J.*, 492, 151842, 2024.
47. Oyetibo, G.O., Enahoro, J.A., Ikwubuzo, C.A., Ukwuoma, C.S., Microbiome of highly polluted coal mine drainage from Onyeama, Nigeria, and its potential for sequestrating toxic heavy metals. *Sci. Rep.*, 11, 1, 17496, 2021.
48. Akoijam, N. and Joshi, S.R., Genome Editing and Genetically Engineered Bacteria for Bioremediation of Heavy Metals, in: *Genome Editing in Bacteria (Part 2),* pp. 184–221, John Wiley & Sons, Hoboken, New Jersey, USA, 2024.
49. Rajoria, S., Vashishtha, M., Sangal, V.K., Electroplating wastewater treatment containing heavy metals and COD using batch and once through continuous electro-oxidation process. *Sep. Sci. Technol.*, 59, 15, 1425–1453, 2024.
50. Ismail, W.W., Syah, M.I.A.I., Muhet, N.H.A., Bakar, N.A., Yusop, H.M., Samah, N.A., Adsorption behavior of heavy metal ions by hybrid inulin-TEOS for water treatment. *Civ. Eng. J.*, 8, 9, 1787–1798, 2022.

51. García, L., Leyva-Díaz, J.C., Díaz, E., Ordóñez, S., A review of the adsorption-biological hybrid processes for the abatement of emerging pollutants: Removal efficiencies, physicochemical analysis, and economic evaluation. *Sci. Total Environ.*, 780, 146554, 2021.
52. Chaudharı, M., Chotalıya, R., Ali, G.H., Pandya, A., Shrivastav, P., Assessment of heavy metal contamination in the groundwater of Gujarat, India using the Heavy Metal Pollution Index. *Environ. Res. Technol.*, 7, 3, 471–488, 2024.
53. Bhardwaj, L.K., Rath, P., Jain, H., Choudhury, M., Exploring the effects of e-waste on soil, water quality and human health. *Discov. Civ. Eng.*, 2, 1, 12, 2025.
54. Bhardwaj, L.K., Rath, P., Choudhury, M., Metzdorff, A., Jodder, P.K., Electronic wastes and contamination of water, in: *Electronic Waste*, pp. 48–57, IntechOpen, London, United Kingdom, 2025.
55. Bhardwaj, L.K., Rath, P., Tokas, R., E-waste management in developing countries: current practices, challenges, disposal, and impact on human health and environment, in: *Sustainable solutions for e-waste and development*, pp. 90–102, Taylor & Francis, London, United Kingdom, 2024.
56. Kumar, V., Dwivedi, S.K., Oh, S., A critical review on lead removal from industrial wastewater: Recent advances and future outlook. *J. Water Process Eng.*, 45, 102518, 2022.
57. Kumaresan, M., Vijayan, A., Ramkumar, M., Philip, N.E., Unraveling the enigma: chronic kidney disease of unknown etiology and its causative factors with a specific focus on dissolved organic compounds in groundwater—reviews and future prospects. *Environ. Geochem. Health*, 46, 12, 510, 2024.
58. Güdek-Gölçek, Ş., Orchestrating Change: The United Nations' Role in Harmonizing Sustainable Development With Global Governance, in: *Harmonizing Global Efforts in Meeting Sustainable Development Goals*, pp. 1–16, MDPI, Basel, Switzerland, 2024.
59. Shittu, E., Lakhanpaul, M., Vigurs, C., Sarkar, K., Koch, M., Parikh, P., Campos, L.C., A rapid systematic scoping review of research on the impacts of water contaminated by chemicals on very young children. *Sci. Total Environ.*, 891, 164604, 2023.
60. Nguyen, Y.T.H. and Nguyen, D.P., The efforts to respond to climate change and implementation of the Sustainable Development Goals (SDGS) from the hardest-affected countries: Vietnam case analysis. *Braz. J. Int. Law*, 19, 165, 2022.
61. Tudor, T. (Ed.), *Resources Management: Global Perspectives and Initiatives*, MDPI, Basel, Switzerland, 2025.
62. Mukura, G.J., Dubovyk, O., Muthoni, F.K., Remote sensing-based land degradation assessment to inform Sustainable Development Goal 15.3: A case study in Ghana, West Africa, 2023.
63. Jin, C.Z., Zhuo, Y., Jin, F.J., Lee, H.G., Jin, L., Genetic diversity into a novel free-living species of Bradyrhizobium from, in: *Microbial OMICS, an asset*

to accelerate sustainability in agricultural and environmental microbiology, Elsevier, Amsterdam, Netherlands, 2024.
64. Ruvinga, C., SMSCS-O-7. *PRIS*, 74, 2024.
65. Correia, J., Leong, S., Sinay, J., Transforming ASEAN for Sustainability: The Role of Regional Integration in Advancing ASEAN's Green Transition, 2025.
66. Kowalska, A. and Biczak, R., Phytoremediation and environmental law: harnessing biomass and microbes to restore soils and advance biofuel innovation. *Energies*, 18, 7, 1860, 2025.
67. Rajput, V.D. and Minkina, T.M. (Eds.), *Nanomaterials and Nano-biochar in Reducing Soil Stress: An Integrated Approach for Sustainable Agriculture*, CRC Press, Boca Raton, Florida, USA, 2024.
68. Nguyen, B.T.N., Kansal, M.L., Nguyen, H.H., Soil Quality Assessment Towards its Sustainable Management in Thai Binh Mangrove, Vietnam. *Water Air Soil Pollut.*, 235, 8, 479, 2024.
69. Shah, A. and Gokhale, D., Navigating micropollutant contamination: a review of industrial wastewater technologies and policies in the US. *MIT Sci. Pol. Rev.*, 5, 30–40, 2024.
70. Alomari, M.A. and Heffron, R.J., Middle eastern energy and climate ambitions: Vision 2030 in the Kingdom of Saudi Arabia and the influence of the legal environmental principles. *J. World Energy Law Bus.*, 16, 5, 442–460, 2023.
71. Chaturvedi, V., Ghosh, A., Garg, A., Avashia, V., Vishwanathan, S.S., Gupta, D., Sinha, N.K., Bhushan, C., Banerjee, S., Datt, D., Bansal, J., India's pathway to net zero by 2070: status, challenges, and way forward. *Environ. Res. Lett.*, 19, 11, 112501, 2024.
72. Sekhohola-Dlamini, L.M., Keshinro, O.M., Masudi, W.L., Cowan, A.K., Elaboration of a phytoremediation strategy for successful and sustainable rehabilitation of disturbed and degraded land. *Minerals*, 12, 2, 111, 2022.
73. Le, T.H., Do, V.K.H., Nguyen, P.H., Do, P.H., Nguyen, T.Q.M., Luu, M.L., Nguyen, T.H., Assessment of Free Chlorine Residual and Microbiological Parameters in Supply Water of Centralized Supply Water Companies in Vietnam. *JST: Eng. Technol. Sustainable Dev.*, 33, 5, 34–41, 2023.
74. Mena-Quintana, F.N., Álvarez, W., Franco, W., Moncayo, L., Tipán, M., Ayala, J., Land Degraded by Gold Mining in the Ecuadorian Amazon: A Proposal for Boosting Ecosystem Restoration Through Induced Revegetation. *Forests*, 16, 2, 372, 2025.
75. Perera, K., Ziajahromi, S., Nash, S.B., Leusch, F.D., Microplastics in Australian indoor air: abundance, characteristics, and implications for human exposure. *Sci. Total Environ.*, 889, 164292, 2023.
76. Kandpal, P.C., *Environmental policy, governance and politics: a South Asian perspective*, CRC Press, Boca Raton, Florida, USA, 2024.
77. Nguyen, T.Q., Nguyen, H.X., Bui, M.Q., Pham, D.H., Truong, H.M., Nguyen, T.N., Hoang, T.M., Truong, M.N., Towards a Standardized Design Framework

for Phytoremediation of contaminants in water: Bridging Research and Policy-Driven Applications, 2025.

78. Ginzky, H., Kameri-Mbote, P., Ruppel, O.C., Sambo, P.T., Tamasang, C.F., Mapping out options for model legislation for sustainable soil management in Africa, in: *African Soil Protection Law*, pp. 379–452, Springer Nature, Singapore, 2021.
79. Berezovyi, V. and Kharchenko, V., Implementation of sustainable development goal 6 in Ukraine: challenges and obstacles (as example of river Ros'), in: *European Dimensions Of Sustainable Development*, 2024.
80. Ogude, J. and Mushonga, T. (Eds.), *Environmental humanities of extraction in Africa: Poetics and politics of exploitation*, Springer Nature, Singapore, 2022.
81. Sharma, H.K., Kumar, A., Singh, V., Meena, H., Meena, B., Sharma, P., Rai, P., Genetic resources of Brassicas, in: *Cash Crops*, Springer Nature, Singapore, 2022.
82. Rugji, J., Erol, Z., Taşçı, F., Musa, L., Hamadani, A., Gündemir, M.G., Karalliu, E., Siddiqui, S.A., Utilization of AI–reshaping the future of food safety, agriculture and food security–a critical review. *Crit. Rev. Food Sci. Nutr.*, 1–45, 2024.

12

Innovating with Nature and Nanotech: Multidisciplinary Case Investigations into Sustainable Heavy Metal Remediation Techniques

R. Venkatesh[1]*, K. Subramani[1] and Laxmi Kant Bhardwaj[2]

[1]Department of Physics, PSNA College of Engineering and Technology, Dindigul, Tamil Nadu, India
[2]Amity Institute of Environmental Toxicology, Safety and Management, Amity University, Noida, Uttar Pradesh, India

Abstract
This chapter examines a collection of multidisciplinary case studies demonstrating the confluence of nature-based solutions and nanotechnology (NT) to achieve sustainable remediation of heavy metal (HM) pollution. The cases examined are located in diverse geographical and environmental contexts and showcase creative strategies for addressing one of the most enduring environmental issues: pollution of soil, water, and air by hazardous HMs. In this chapter, case studies like phytoremediation with nanomaterials (NMs), sediment remediation using microbial biosynthesis of nanoparticles (NPs), and biochar-nanocomposite filters for urban water, reveal the potential of techniques inspired by nature and the application of innovative NTs to enhance remediation potential significantly. The chapter points to examples of collaboration between environmental scientists, materials scientists, and biotechnologists to develop approaches that are practical, resource-efficient, and environmentally sustainable. The case studies were provided to highlight the solutions needed for local pollution problems, and under specific environmental conditions. Finally, the chapter highlights the facilitative role that NT-inspired innovation plays in augmenting the effectiveness of classical approaches, i.e., phytoremediation, through enhanced metal bioavailability or enhanced contaminant

**Corresponding author*: sukatesh@gmail.com; ORCID: https://orcid.org/0000-0002-0355-8857

Prangya Rath, Laxmi Kant Bhardwaj and Amit Kumar (eds.) Heavy Metal Remediation, (285–310)

adsorption capacity. This chapter importantly outlines lessons in the scalability, environmental friendliness, and long-term sustainability of these technologies through comparative scrutiny of their trials and successes.

Keywords: Case studies, environmental sustainability, heavy metal remediation, nanotechnology, nature-based solutions, phytoremediation

12.1 Introduction to Sustainable Heavy Metal (HM) Remediation

Heavy metal (HM) contamination is still posing one of the greatest environmental concerns worldwide, especially in highly dense locales undergoing modifications due to modernization, mining, and agricultural runoff. HM, such as arsenic (As), lead (Pb), cadmium (Cd), and mercury (Hg), pose valiantly diversified situations that threaten families as well as ecosystems [1, 2]. These contaminants linger in the environment and, in some cases, such as for a certain HM, remain there for decades or more. This is largely because HMs are non-biodegradable environmental contaminants and are often highly bioaccumulative along the food chain, causing concern for the succeeding generations and public health way beyond major clean-up operations [3]. Illnesses caused by HMs include neurological disorders, cancers, and other degenerative implications on the public health infrastructure. Hence, it is quintessential that contaminated sites where HMs are involved should be remediated toward sustainability.

Traditional remediation pathways, such as physical removal and chemical treatment, have been the predominant methods for treating HMs. However, traditional methods carry high costs and energy requirements, produce secondary waste, and contaminating a larger or more complex site may be impractical. In addition, there is a growing focus on low-cost and ecologically acceptable remediation methods that provide sustainable long-term approaches to HM pollution cleanups [4]. Table 12.1 summarizes HM contamination, the sources, environmental and health impacts, and remediation options. It notes the change from traditional approaches to sustainable, cost-effective options.

In recent years, the integration of nature-based solutions and NT has presented a promising path forward for the remediation of HMs (see Figure 12.1). Nature-based solutions utilize natural processes, organisms, and systems to restore the environment and manage pollution. Phytoremediation might be thought of as the use of plants to absorb and concentrate HMs from systemically contaminated (soils or water), while bioremediation

Table 12.1 Overview of heavy metal (HM) contamination: Sources, impacts, and remediation strategies.

S. no.	Aspect	Details
1	Heavy Metals (HMs)	Arsenic (As), Lead (Pb), Cadmium (Cd) and Mercury (Hg)
2	Common Sources	Industrial effluents, Mining and smelting, Fertilizers and pesticides, Improper waste disposal
3	Typical Concentrations in Polluted Soil	Lead (Pb): up to 5,000.0 mg/kg; Mercury (Hg): 0.1–10.0 mg/kg; Cadmium (Cd): 1.0–1,000.0 mg/kg; Arsenic (As): 10.0–3,000.0 mg/kg
4	Environmental Persistence	Non-biodegradable, Half-lives often exceed 100 years, Depending on the metal type and conditions
5	Bioaccumulation Factor (BAF)	Cadmium (Cd) in plants: 10–1,000 Mercury (Hg) in fish: >10,000 (especially as methylmercury)
6	Health Impacts	Lead (Pb): Reduced Intelligence Quotient (IQ), and kidney damage, Mercury (Hg): Neurotoxicity and fetal development issues, Cadmium (Cd): Kidney, lung, and bone damage Arsenic (As): Cancers
7	Ecological Impacts	Soil degradation, Reduced microbial activity, Aquatic toxicity: fish mortality and biodiversity loss
8	Traditional Remediation Methods	Soil excavation, Chemical stabilization (e.g., lime, phosphates), Soil washing

(Continued)

Table 12.1 Overview of heavy metal (HM) contamination: Sources, impacts, and remediation strategies. (*Continued*)

S. no.	Aspect	Details
9	Limitations of Traditional Methods	High cost: \$100 to \$1,000 per ton High energy demand Secondary waste production Unsuitable for large or diffuse sites
10	Emerging Sustainable Techniques	Phytoremediation, Bioremediation (bacteria/fungi), Electrokinetic remediation, Biochar application
11	Advantages of Sustainable Methods	Lower cost: \$20 to \$100 per ton (e.g., phytoremediation), Low secondary pollution, Scalable and community-friendly
12	Global Concern Level	Over 10 million contaminated sites worldwide, More than 140 million people affected (World Health Organization (WHO) estimates)
13	Policy and Regulatory Drivers	Guidelines by the European Union (EU), Environmental Protection Agency (EPA), World Health Organization (WHO), and the Soil Directive Permissible soil limits: Lead (Pb) (<300.0 mg/kg), Arsenic (As) (<20.0 mg/kg), and Cadmium (Cd) (<3.0 mg/kg)

would be the use of microorganisms to degrade or transform pollutants to less harmful forms [5]. Nature-based solutions can be advantageous because of their tendency to be less expensive, environmentally friendly, and can address larger areas with little to no secondary waste [6].

Nanotechnology (NT) employs halting the material at the nanoscale, usually between 1 and 100 nanometers, creating new materials with desired attributes. One of the advantages of NT in HM remediation is that it can increase the efficacy and selectivity of natural approaches. Soil and water pollution can be countered either by adsorption or chemical transformation of HMs using naturally synthesized NPs (examples: zero-valent iron

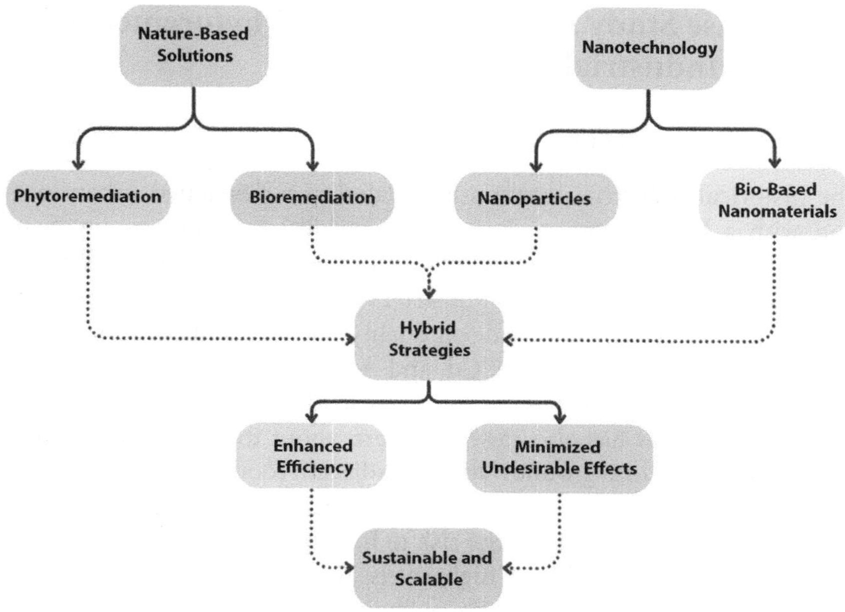

Figure 12.1 Hybrid strategies for heavy metal (HM) remediation.

NPs, metal oxide NPs). Thus, because of sustainability and biodegradability, bio-NMs (like plant frass or chitin) would be preferable. Many new hybrid approaches have emerged that highlight the benefits of a natural approach that incorporates the use of NT to increase efficacy and reduce unwanted impacts. This presents more sustainable and scalable approaches than conventional methods, and can allow for more effective cleaning of problematic environments.

This chapter reviews several multidisciplinary case studies that demonstrate real-life applications of nature-based and nanotech-enhanced solutions for the remediation of HM-contaminated sites. The case studies cover different locations, types of contamination, and types of remediation approaches, and demonstrate the versatility and effectiveness of these solutions in real circumstances. Each case study provides a unique example that integrates natural systems and NT in addressing specific environmental problems. These case studies are established based on technology type (e.g., phytoremediation with NPs, biosynthesized NPs for water treatment, biochar-nanocomposite filters for wastewater), based on the matrix environment (soil, water, or sediment), and the scale of the remediation.

12.2 Case Study 1: Nano-Enhanced Phytoremediation in Industrial Soil

12.2.1 Site Profile and Contaminants

The historical industrial location is situated in a developed urban location with a long history of heavily industrialized operations involving massive amounts of heavy manufacturing, metal processing, and chemical processes. Over the years, industrial activities at the site have deeply contaminated adjacent soils and soil quality due to cumulative impacts of leaching HMs, in particular Pb, Cd, and As, into soils that may be affecting the health of ecosystems and human health. The location's proximity to residential areas has also raised concern about the possibility of contaminants getting into groundwater and adjacent food crops that may also be contaminated, thus creating an environmental remediation opportunity for the facility to mitigate the risk to human health [7]. While the site is contaminated with HMs at high levels, they have endured over time, making remediation options such as chemical treatment and excavation costly, ineffective, and unsustainable. Ultimately, the potentially costly bulk removal of soil and the long-term impact of HMs being left on site led to the development of a sustainable, cost-effective approach to restore the site using nano-enhanced phytoremediation techniques. Table 12.2 gives a comprehensive characterization of the polluted industrial site, mentioning pollutant sources, related environmental risks, and justification for picking the nano-enhanced phytoremediation technique.

12.2.2 Methodology: Nanoparticle (NP) Assisted Hyperaccumulators

Soil remediation through phytoremediation relies on plants that absorb pollutants and break down contaminants while being recognized as an environmentally sustainable approach (Figure 12.2). Traditional phytoremediation effectiveness faces restrictions because of its limited uptake capacity and its gradual remediation process. In order to address these constraints through novel phytoremediation methods, researchers involved in this case study turned to NT, specifically through the use of engineered NPs to enhance the conventional phytoremediation method [8].

The main difference with this case study is that they used nano-sized particles (specifically metal oxide NPs such as zinc oxide (ZnO) and iron oxide (Fe_3O_4)) to enhance the ability of plants to uptake and accumulate

Table 12.2 Site overview and remediation context nano-enhanced phytoremediation case study.

S. no.	Category	Details
1	Site Location	Urbanized industrial area with a long history of heavy manufacturing, metal processing, and chemical production
2	Primary Contaminants	Lead (Pb), Cadmium (Cd), and Arsenic (As)
3	Contamination Source	Decades of industrial activities have led to the leaching of heavy metals (HMs) into the surrounding soil
4	Environmental Concerns	Soil quality degradation, Risk of heavy metals (HMs) contaminating groundwater and nearby food crops, Ecosystem and human health risks
5	Proximity to Population	Close to residential areas, which increases the risk of exposure to contaminants
6	Challenges in Remediation	High contamination levels, Widespread contamination across a large area, Chemical treatment and excavation are costly and unsustainable
7	Rationale for Nano-Enhanced Phytoremediation	Conventional methods are ineffective and expensive, Need for a scalable, cost-effective, and sustainable solution
8	Selected Approach	Nano-enhanced phytoremediation – combining nanotechnology (NT) with plant-based remediation to extract or stabilize heavy metals (HMs) in soil
9	Objectives of Remediation	Reduce heavy metal (HM) concentration, Minimize human and environmental health risks, Restore soil quality sustainably

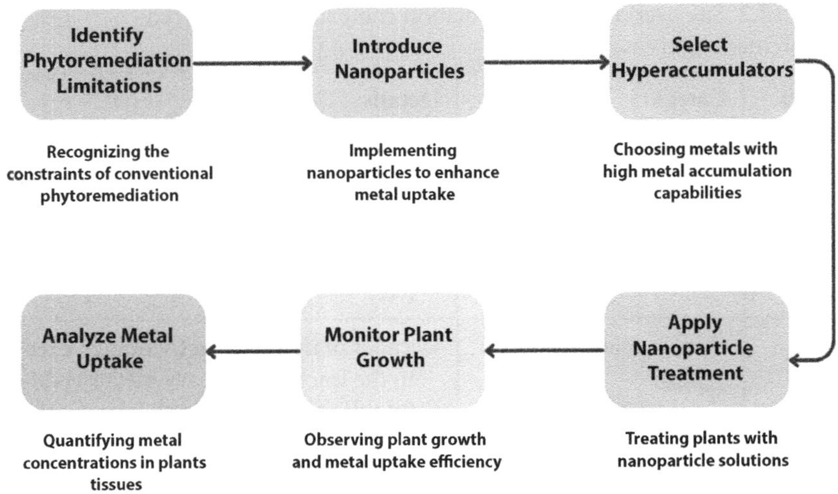

Figure 12.2 Nanoparticle (NP) assisted with the phytoremediation process.

HMs from contaminated soil. The NPs were selected because they could interact with plant roots, which increased the surface area for metal absorption, while increasing plants' tolerance to toxicity. These NPs were synthesized to be nontoxic to the plants but highly reactive towards the intended target contaminants, allowing the plants to absorb metals far more effectively. The plants chosen for this study were hyperaccumulators, essentially plants that have the ability to take up and sequester elevated concentrations of metals in their tissues. Commonly used hyperaccumulators in this study were the Brassica varieties (mustards) and varieties of sunflower [9]. In this study, the plants were treated with NP solutions through either soaking the roots directly in the solutions or by actually treating the plants with foliar sprays. A principal aspect of this methodology was a focused examination of plant growth and tracking metal uptake efficiencies, in addition to measuring negative effects that may be caused by the NPs themselves. The growing media were analyzed for metal concentrations at the beginning and end of the treatment period, and the plants were harvested so that the metals could be quantified in the tissue of various organs (roots, stems, leaves).

12.2.3 Outcomes and Challenges

The outcomes of the nano-enhanced phytoremediation process were positive. The plants treated with NPs took up significantly higher concentrations of metals than the untreated plants. In expanding upon the results themselves, ZnO-NPs supported brassica species to improve their ability to accumulate Pb and Cd in their roots, and Fe_3O_4-NPs supported the uptake of As by sunflower plants. As a result, the number of HMs in the soil was reduced by 50–70% during the time frame, which was typically around 3–6 months for the remediation process. The plants cultured using the NPs also exhibited increased resistance to metal-induced toxicity, which is usually the biggest problem encountered within the phytoremediation process. Untreated hyperaccumulators that are cultured in HM loads tend to exhibit stunted growth and increased physiological stress, but the incorporation of NPs was able to remove these stresses, enabling the plant to grow healthily and produce biomass. Moreover, having healthy plants shows increased effectiveness for the outcomes from both survivable yield and biomass production for the phytoremediation process [10]. Furthermore, incorporating NT within the phytoremediation process allowed for a more efficient and more scalable operation than was possible with conventional treatments. The higher metal uptake capacity and faster remediation timeframes reduced the need for a continuous or intensive monitoring approach, especially for profit-based purposes.

Although there were some exciting results, there were also challenges encountered throughout this study. One of the main issues was the possible toxicity of NPs to non-target organisms in the soil. The NPs under investigation were formulated to be safe for the plants; however, there was the possibility of targeting soil microorganisms, earthworms, and other beneficial organisms that contribute to the health of the soil. NPs can aggregate in the exposed environment or become immobilized, both of which would reduce the intended benefit of improving metal uptake. The study suggested that the NPs were available in soil for many months; however, further studies would be needed to determine the long-term persistence and behavior of these materials in the environment, including upstream interactions with the groundwater (either through leaching or accumulation). The method was successful in a controlled field environment; however, scaling to larger sizes with multiple competing environmental conditions (e.g., soil type, degree of contamination) would require further optimization. Additionally, the cost of synthesizing the NPs, in addition to the potential need for larger quantities of NPs, may pose general economic obstacles for large-scale applications. The nano-enhanced phytoremediation approach was shown to be a promising measure for remediating industrial soils contaminated with HMs. The NPs integrated with hyperaccumulator plants improved metal uptake potential, reduced plant toxicity, and generally improved soil decontamination rates.

12.3 Case Study 2: Biosynthesized Nanoparticles (NPs) for River Sediment Clean-Up

12.3.1 Problem Context: Mining-Related Heavy Metal (HM) Discharge

River systems are frequently subject to pollution as they are adjacent to industrial and mining activities that often contribute HMs to water and sediment. This case study involved a waterway through an area that had previously been occupied by metal mining operations. Due to significant unregulated small-scale mining, sediments in this river were contaminated with HMs like Hg, As, and copper (Cu). Sediments contaminated with HMs and other pollutants can pose risks for many years to both local aquatic ecosystems and people living near the waterway who are using the water for drinking, irrigation, or recreation [11]. In the context of this study, the offending activity was gold (Au) mining, with mining generally restricting the concentration of Au (in the ore) using Hg. The Hg resulted in increased levels of Hg in the sediment and water of the river, and it was then bioaccumulated in the aquatic organisms, ultimately causing a potential risk to the wildlife and humans that received it through their food [12]. With these challenges in mind, a more sustainable and innovative approach was necessary. This case study describes the application of microbial biosynthesized NPs as an Hg-contaminated river sediment remediation technique. The NPs were manufactured to bind to the Hg and HMs found in the sediments, which would effectively immobilize both the Hg and HMs and prevent further contamination of the water column [13]. Table 12.3 concerns contamination in sediments, with its contextualization taking place in a riverine system as affected by unregulated Au mining.

12.3.2 Microbial Synthesis of Remediation Agents

The unique and innovative aspect of this research is leveraging microbial synthesis to form NPs targeting HM remediation. The research focused on an organism such as *Bacillus* and *Pseudomonas* with natural metal-reducing capabilities, particularly identifying that they can reduce and precipitate HMs. The microorganisms were cultivated in the lab to generate stainless metal oxide NPs, Fe_3O_4-NPs, ZnO-NPs, etc. The metal oxide NPs were of particular importance, as the generated NPs were good at adsorbing and binding Hg ions from the sediment, which immobilizes the Hg, preventing further contaminations [14]. The process began with collecting river

Table 12.3 River sediment remediation using biosynthesized nanoparticles (NPs), heavy metal (HM) contamination from mining.

S. no.	Category	Details
1	Site Location	River adjacent to a former gold (Au) mining area, affected by unregulated small-scale mining
2	Primary Contaminants	Mercury (Hg), Arsenic (As), and Copper (Cu)
3	Contamination Source	Discharge from small-scale gold (Au) mining using mercury (Hg) for gold (Au) extraction, leading to metal accumulation in river sediment
4	Environmental Concerns	Long-term pollution of sediment, Bioaccumulation of mercury (Hg) in aquatic organisms, Human exposure *via* the food chain and water use
5	Affected Communities	Populations use the river for drinking water, irrigation, and recreation
6	Challenges in Remediation	Standard methods like dredging and chemical treatment were costly, invasive, and environmentally disruptive
7	Rationale for Innovative Approach	Need for a low-impact, cost-effective alternative, Desire to prevent further contamination while preserving ecosystem integrity
8	Selected Approach	Use of microbial biosynthesized nanoparticles (NPs) to immobilize mercury (Hg) and heavy metals (HMs) in the sediment
9	Mechanism of Action	Nanoparticles (NPs) bind to contaminants, immobilizing them and reducing their mobility and bioavailability in the sediment
10	Objectives of Remediation	Reduce the risk of heavy metal (HM) exposure to humans and wildlife, Stabilize contaminated sediment, Prevent further spread of pollutants

sediment from several heavily contaminated locations (Figure 12.3). The sediment was added to the microbial strains in biosafety cabinets, under identified controlled conditions. The microorganisms were allowed to synthesize NPs when exposed to Hg; the NPs formed represent the products of microbial enzymatic reduction of Hg ions to a less toxic immobilized form,

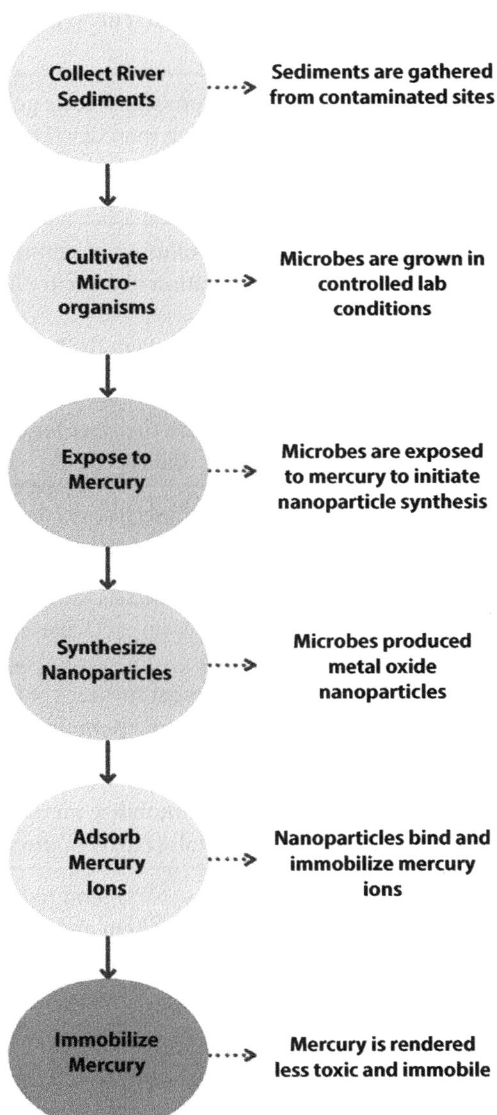

Figure 12.3 Microbial synthesis of nanoparticles (NPs) for heavy metal (HM) remediation.

while also producing a stable interaction of HMs with the NPs. The microbial biosynthetic method has several advantages over chemical processes. They were more environmentally friendly, less expensive, and sustainable as they employed local microbes that did not need the use of harmful chemicals. Another advantage the biosynthesized NPs offered was a reduction in the introduction of additional pollution. The biosynthetic approach is a compelling option for river sediment remediation projects [15].

12.3.3 Results and Environmental Performance

The results of the biosynthesis and application of the NPs to the contaminated river sediments were encouraging. After adding microbial cultures and allowing enough time for NPs to form, biosynthesized NPs were then added to the sediment due to their potential to reduce Hg from the sediment. The research results provide important details about the value of biosynthesized NPs as a way to remediate river sediments that have been contaminated with Hg. More specifically, Fe_3O_4-NPs produced a 70–80% reduction in Hg concentration over a two-month period, whereas ZnO-NPs produced a noticeable reduction but at approximately 50–60% effectiveness. Nonetheless, the NPs immobilized metal ions and prevented the ongoing migration of Hg from sediment into the water column. Treatment with NPs not only functions to reduce Hg but also produces signs of biological enhancement in treated sediments, and results in better conditions for indigenous microorganisms. Along with the highly reduced Hg concentrations in the river sediments, after treatment, downstream water quality measurements showed continued signs of Hg reduction in water column samples taken after reclamation that contained reduced Hg than samples taken before reclamation. Thus, this indicates that the Hg that was bound in sediments was not being re-released back into the river. Treatment with NPs also imparted important long-term stability as NPs remained active long after the treatment process was completed and did not leach off Hg, suggesting that these NPs formed stable bonds with Hg, sequestering it, and preventing future triad effects that can be observed in the environmental system.

While promising, a number of challenges arose in the use of biosynthesized NPs in cleaning river sediment. One of the principal issues is scalability; while the microbial synthesis was successful in laboratory and small-scale field tests, it becomes seriously problematic to scale up production for cleaning contamination over larger river systems. The large number of NPs needed may implement on a large scale prohibitively expensive, and further research into the optimization of microbial NP production is warranted. Furthermore, although no short-term adverse effects were seen in sediment biology, long-term consequences to non-target organisms such as fish and aquatic plants are unknown and need investigation. A further essential challenge is the absence of formal environmental regulations to govern the application of NPs to the process of remediation. This emphasizes the need for policy development to ensure that NTs can be properly and safely implemented for the cleansing of the environment. But biosynthesized NPs provide a greener, eco-friendly alternative for HMs remediation, e.g., Hg, particularly in areas damaged by poisonous mining activities. Their green, low-cost, and versatility render them an appealing solution for contamination control, subject to site conditions and levels of environmental degradation [16].

12.4 Case Study 3: Biochar-Nanocomposite Filters in Urban Water Systems

12.4.1 Urban Wastewater and Metal Contamination

Urban water systems are subject to a wide range of problems due to industrial and domestic pollution. Urban wastewater, particularly industrial wastewater, is frequently contaminated with a lot of hazardous substances: HMs such as Pb, Cd, chromium (Cr), and Cu. These pollutants present critical risks to human health, aquatic life, and the overall quality of water. Pollutants are toxic, non-biodegradable, and have a tendency to bioaccumulate [17]. Traditional wastewater treatment, including the conventional treatment of wastewater *via* filtration and chemical treatments, is not always effectively capable of removing HMs and is often not a cost-effective solution for large-scale projects. For this case study, the intention was to test whether biochar-nanocomposite filters could be an effective and sustainable technology to treat HM-contaminated urban wastewater. Biochar, a carbon-rich material produced from renewable organic waste, is coupled with NMs (Fe_3O_4 and carbon-based NPs) as a contributive solution to increase HM removal effectiveness from contaminated wastewater. Table 12.4 illustrates HM contamination problems in urban water systems while focusing on the biochar-nanocomposite filter for a sustainable treatment solution.

12.4.2 Filter Design and Deployment

In this case study, the biochar-nanocomposite filter contains agricultural waste biochar (i.e., rice husk and coconut shell) and engineered NPs

Table 12.4 Application of biochar-nanocomposite filters in urban wastewater treatment.

S. no.	Category	Details
1	Site Location	An urban area with significant industrial and domestic wastewater discharge
2	Primary Contaminants	Lead (Pb), cadmium (Cd), chromium (Cr), and copper (Cu)
3	Contamination Source	Wastewater from industrial zones and domestic sewage contains heavy metals (HMs)
4	Environmental Concerns	Toxicity and bioaccumulation of metals, Threats to drinking water quality aquatic ecosystem degradation
5	Challenges in Remediation	Traditional wastewater treatment is often ineffective against heavy metals (HMs), High cost and limited scalability of existing solutions
6	Rationale for Innovative Approach	Need for an efficient, low-cost, and sustainable filtration method, Adaptation to increasing contamination from urbanization
7	Selected Approach	Use of biochar-nanocomposite filters for heavy metal (HM) removal from urban wastewater
8	Filter Composition	Biochar derived from organic waste combined with nanomaterials (NMs) (e.g., iron oxide (Fe_3O_4), carbon-based nanoparticles (NPs))
9	Mechanism of Action	Adsorption of heavy metals (HMs) onto the biochar-nano surface, enhancing removal efficiency and reducing toxicity
10	Objectives of Remediation	Improve the removal of heavy metals (HMs) from wastewater, Offer a scalable, eco-friendly solution, Protect water quality and ecosystems

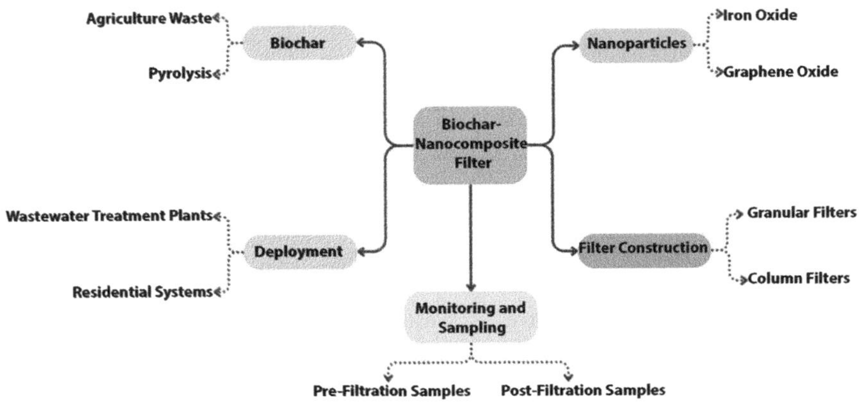

Figure 12.4 Biochar-nanocomposite filter design and deployment.

(Figure 12.4). Due to its high surface area, porosity, and adsorption capabilities, biochar is a good adsorbent for contaminants. However, its HM adsorption capacity was enhanced by using reactive NPs for the modifications (e.g., using Fe_3O_4-NPs and graphene oxide [18]. The NP modification approach was conducted by impregnating biochar with an NP solution. The NPs coated the surface of the biochar and occupied the pore volume. This, in turn, improved the sorptive capacity of the biochar as well as chemically reduced and immobilized metals like Pb and Cd. The final biochar-nanocomposite end product was lightweight and robust, which can be directly deployed to filtration systems [19]. The filters were crafted in various formats, including granular and column types, to fit the requirements of different urban water systems. The filters were used in pilot-scale urban water treatment plants and residential wastewater treatment systems, targeting areas with higher amounts of contamination. Water samples were taken pre- and post-filtration to characterize the success of the biochar-nanocomposite filters in removing HMs from wastewater.

- ***Key Methods Implemented During the Filter Operation Include***

The major techniques utilized during the process of filtering included a step-by-step method to design and deploy an efficient water treatment system through the use of NT-boosted materials. To begin with, biochar was generated through pyrolysis and then functionalized with NPs to create a biochar-nanocomposite with improved adsorption capacity. This altered material was then utilized to build filters, which were loaded into pre-fabricated units designed for the application size, either granular media or in columnar shape. These filters were installed in water treatment plants or community-level water treatment plants, where water was filtered through the filters for the removal of contaminants. For the efficiency of the system, routine monitoring and sampling of water throughout all the phases of filtration were conducted, permitting rigorous analysis of the treatment process and providing insights into the operational efficiency of the biochar-nanocomposite filters.

12.4.3 Evaluation of Efficiency and Reusability

The efficiency of the biochar-nanocomposite filters to remove HMs from urban wastewater was assessed through several parameters, including the removal efficiency, filter longevity, and reusability. The biochar-nanocomposite filters were found to have excellent efficiency in removing various HMs from polluted water, far surpassing typical activated carbon filters. In particular, the filters removed Pb at 85–90%, successfully reducing levels below drinking water limits. Cd removal was slightly less efficient at 75–80%, but still showed excellent performance. Cu and Cr were also treated efficiently, with removal efficiencies of 80–85% and 70–75%, respectively. The enhanced performance of the above two filters can be explained on the basis of the addition of NPs, which added reactivity and adsorptive capacity to the biochar. Apart from physical adsorption, the NPs also provided for chemical reduction processes that further contributed to the enhanced removal efficiency of some of the HMs [20]. These findings indicate that biochar-nanocomposite filters provide a very effective and innovative water treatment solution for HM remediation.

The biochar-nanocomposite filter system had many qualities that made it advantageous, and the potential to regenerate and reuse the filter media was one of them. The usage length of the filters was determined by their effectiveness in removing contaminants before the adsorption capacity was sufficiently degraded. Once the filters had reached saturation, they were regenerated by washing them out with a mild acid solution (e.g., citric acid), which effectively desorbed the HMs. Subsequently, the filters had a good amount of their adsorption capacity regained, and could be reused over many cycles. It was determined that the filters have the potential to be reused for approximately 10 cycles without significant losses in efficiency. After 10 cycles, a decline in removal efficiency was evident; however, regeneration allowed for an extended practical lifespan of the filter. The reusability aspect also added cost-effectiveness to the technology while alleviating some of the environmental impact from the production of new filter media. Filters need to be maintained regularly by periodically regenerating and cleaning to enable continuous performance. The costs associated with regenerating the biochar-nanocomposite filters were low, as the primary material (biochar) was derived from waste, and

NPs were used in small quantities. The biomass-nanocomposite filters were an effective, sustainable, and economical method for removing HMs from urban wastewater systems. The high removal efficiencies, reusability, and relatively low costs of care made them a suitable approach in advancing the quality of water in polluted urban spaces. The biomass-nanocomposite system has great potential, but more work is required to understand and optimize the stability of NPs, and how to scale the process for applicability, as well as the long-term environmental impacts of such a technology [21].

12.4.4 Challenges and Limitations

One concern that could arise is leachate from the biochar during treated water phases, namely, the regeneration process. The NPs used in this study were meant to be stable, though any possibility of leaching during continued use must be monitored to ensure that no secondary contamination occurs. While the pilot scale showed good potential, challenges with scaling the system for larger urban uses or widespread commercial application include production cost, transport, and system design to handle larger volumes of water. In some of the more contaminated areas, the filters approached saturation sooner than others, which meant needing to regenerate, or change the biochar, more often. Finding additional ways to improve the adsorption capacity of the biochar may promote the sustainability of the system overall.

12.5 Cross-Case Synthesis and Lessons Learned

The case studies highlighted throughout this chapter reveal a range of transdisciplinary and innovative approaches for managing HM contamination with regard to sustainability. Although there is variation across these approximations of context, contaminants managed, and technological applications, this synthesis suggests a number of similarities and differences across the approaches and outcomes. Table 12.5 synthesizes the main remediation methods applied in five case studies, contrasting their effectiveness, performance, and usability.

12.5.1 Phytoremediation and Nanotechnology Synergies

Case Studies 1 and 2 demonstrated the synergistic application of phytoremediation with NT (e.g., NPs can be used to enhance metal uptake by plants or immobilize contaminants in soils). In Case Study 1, the use of nano-enhanced phytoremediation of contaminated industrial soil demonstrated

Table 12.5 Cross-case comparison of remediation strategies and outcomes.

S. no.	Approach	Case studies	Outcomes	Effectiveness and insights
1	Phytoremediation and Nanotechnology (NT) Synergies	Case Studies 1, 2	Enhanced metal uptake and immobilization using plant-nano systems. Case 1: Industrial soil remediation showed strong results for nickel (Ni) and copper (Cu). Case 2: Biosynthesized nanoparticles (NPs) used with microbes for river sediment cleanup.	High removal rates for metals such as mercury (Hg), cadmium (Cd), lead (Pb), and nickel (Ni). Effectiveness depends on nanoparticle (NP) synthesis quality, compatibility with the environment, and metal type. Synergy is observed when both biological and technological systems are optimized.
2	Biosynthesized Nanoparticles (NPs) and Biochar-Based Filtration	Case Studies 2, 3	Case 2: Microbial synthesis of nanoparticles (NPs) for sediment remediation. Case 3: Biochar-nanocomposites used in urban wastewater filtration.	Biochar systems demonstrated high filtration and adsorption in urban water treatment—low cost, easy to maintain. Biosynthesized nanoparticles (NPs) are effective, but scalability remains a limitation for broader use.

much greater efficiencies of metal removal, especially nickel (Ni) and Cu, indicating greater synergies between both the plant biology and the function of NMs. Case Study 2 was also another successful example leveraging the microorganisms that were used to synthesize NPs for river sediment remediation, and an example of bioremediation combined with NT for the removal of contaminants from the water system. Both methods exhibited high removal rates, but effectiveness was dependent upon correct NP synthesis, environmental compatibility, and the nature of the HMs.

12.5.2 Biosynthesized Nanoparticles (NPs) and Biochar-Based Filtration

Case Study 3 dealt with biochar-nanocomposite filters for urban wastewater treatment, while in Case Study 2, the biosynthesized NPs were applied to remediate river sediments. Both methods proved to be effective for removing HMs from contaminated water. Biochar-nanocomposites have excellent filtration for urban water open systems, effective metal adsorption, and are easily reused, which are a low-maintenance and sustainable option for continuous water purification. The biochar-based filtration system was particularly effective in urban contexts due to its simple design, scalability, and low cost. Although the NPs produced using bio-synthesis also appeared targeted, they were less scalable for mineralization in water systems.

12.5.3 Success Factors and Limitations

The combination of biological sciences with NT and environmental engineering served as a fundamental success element in every case study. The development of broad and efficient remediation strategies occurred through collaborative approaches. Every case study showed that local environmental factors, including soil properties and water characteristics, along with contamination types, determined how effective the techniques would be. The effectiveness of interventions varied with respect to the components of the other environmental conditions. The pathways of success for phytoremediation and biochar filtration with nanocomposites based on plants were both environmental and cost-effective. The materials are inexpensive because they use plant waste and biochars, and they are reusable, which allows for long-term viability. The success of implementing NMs depended on the use of NMs as NPs in soil and water remediation and included biochar composites. NT increased the adsorption capacity and removal efficiency of the materials while providing targeted remediation of HMs.

The techniques performed well at pilot-scale, but scaling them for implementation at e-waste recycling sites, industrial areas, or agro-ecosystems would present an enormous challenge with cost and logistic constraints, and challenging obstacles presented by the integration of technologies into existing systems. Although they were highly effective, questions were raised about the stability of their NPs and the possible risks of leaching into the environment. This is particularly true for biosynthesized NPs and magnetic NPs, which, if not treated appropriately, can present further environmental risks. Some case studies demonstrated the complexity of multi-source contamination, where several pollutants (HMs, organic compounds, etc.) had to be managed simultaneously. Although in some instances phytoremediation with NT seemed to be effective, in other locations, new and innovative techniques will likely be necessary to manage contaminants in a more integrated approach. Application of sophisticated NMs for environmental remediation purposes, especially in agriculture and drinking water, is bound to have regulatory hurdles or social resistance. Additional research, as well as consultation among local communities, regulators, and policy makers, will need to be done in order to foster general acceptance as well as to guarantee safe application.

12.6 Future Directions and Recommendations

The case studies indicate the promise shown by innovative ideas in HM remediation to improve environmental health. To further increase their activity, an area of research should seek to optimize NT for safer and more stable NPs as well as to develop methods that work to combine plants, microbes, and NMs to mitigate complex contamination in such areas as urban wastes and agriculture. The involvement of local communities, together with the study of social and economic impacts, will guarantee that such solutions are sustainable and accepted by the communities. The suggestions focus on promoting the integration of NT with nature-based solutions to combat HM pollution more efficiently and sustainably. The suggestions ask for the design optimization of NPs for safety, stability, and environmental suitability while being adaptable to local environmental and socio-economic conditions. In addition, the upscaling of effective pilot projects needs to be accompanied by investments in cost-efficient manufacturing and deployment techniques. Strengthening regulatory systems, improving interdisciplinary cooperation, and interacting with local societies are also crucial in order to facilitate acceptance and prudent usage. All implementations should be followed by long-term monitoring and impact evaluation to provide both environmental protection and a sustainable remediation effect.

12.7 Conclusion: Toward Scalable, Nature-Inspired Nanotech Solutions

The examples covered in this study provide a very positive prospect for cleaning up HM pollution as a result of synergistically applying bioremediation with NT. By deploying plant and microorganism (and other bioremediation agents) with recyclable NMs, these approaches are novel, environmentally friendly, and scalable means to address HM contamination of land surfaces and water bodies. Approaches like nano-augmented phytoremediation, biosynthesized NPs, and biochar-nanocomposite filters have shown that nature-inspired solutions expedited by NT are a great leap toward accelerating the effectiveness, affordability, and sustainability of remediation. Such approaches not only remediate and decontaminate but also improve soils, improve groundwater quality, and safeguard public health.

First, it is difficult to extrapolate from small pilot tests to big projects and requires further research. Second, the environmental and social acceptability of NT relies on knowing its long-term effects, stability, and possible risk factors, as well as transparent communication and rules. Third, specialists in multiple areas need to coordinate their efforts to create solutions that respond to local needs and conditions. Last, while implementation of successful practices is paramount, review and refinement must happen to ensure continuity in the long-term impact of ongoing solutions. In conclusion, the future for HM remediation is the use of scalable, nature-based nanotech solutions. While these approaches utilize NP technology to mimic nature's dynamic adaptability, alongside using nature's ability at a scale we can control, which involves optimizing known technologies to radically change environmental cleanup and restoration as we know it. While it seems we are leaps away from perfected and sustainable remediation techniques with significant cost savings in our approaches, and a planet that is cleaner and healthier to live on, it is reasonable to expect that, as we continue to advance, scientifically, the prospects for mitigating HM contamination and ecosystem health are ever brighter.

List of Abbreviations

As	Arsenic
Au	Gold
BAF	Bioaccumulation Factor
Cd	Cadmium
Cr	Chromium
Cu	Copper
EPA	Environmental Protection Agency
EU	European Union
Fe_3O_4	Iron oxide
Hg	Mercury
HMs	Heavy metals
IQ	Intelligence Quotient

Ni	Nickle
NT	Nanotechnology
NMs	Nanomaterials
NPs	Nanoparticles
Pb	Lead
WHO	World Health Organization
ZnO	Zinc oxide

AI disclosure statement: The author acknowledges the use of AI-based tools for minor language editing or grammatical correction. However, all intellectual contributions, analyses, and conclusions presented in this chapter are entirely the author's own.

References

1. Gupta, U., Shukla, S., Bhardwaj, L.K., Rath, P., Upadhyay, D., Sharma, B., Rupam, P., Asthana, H., Jindal, T., Singh, H., Unraveling Urban Plant Strategies for Heavy Metal Uptake and Detoxification, in: *Urban Forests, Climate Change and Environmental Pollution*, pp. 93–119, Springer, Cham, 2024, https://doi.org/10.1007/978-3-031-67837-0_5.
2. Alam, S., Bhardwaj, L.K., Mallick, R., Rai, S., Estimation of heavy metals and fluoride ion in vegetables grown nearby the stretch of River Yamuna, Delhi (NCR), India. *Indian J. Environ. Prot.*, 43, 1, 64–73, 2023.
3. Selvasembian, R., Thokchom, B., Singh, P., Jawad, A.H., Gwenzi, W. (Eds.), *Remediation of Heavy Metals: Sustainable Technologies and Recent Advances*, Wiley, 2023, https://doi.org/10.1002/9781119853589.
4. Sharma, S., Tiwari, S., Hasan, A., Saxena, V., Pandey, L.M., Recent advances in conventional and contemporary methods for remediation of heavy metal-contaminated soils. *3 Biotech*, 8, 4, 2018, https://doi.org/10.1007/s13205-018-1237-8.
5. Bhardwaj, L.K., Kumar, D., Kumar, A., Phytoremediation Potential of Ocimum sanctum, in: *Phytoremediation Potential of Medicinal and Aromatic Plants*, pp. 46–57, CRC Press, 2023, https://doi.org/10.1201/9781003290780-5.
6. Adeoba, M.I., Odjegba, E.E., Pandelani, T., Nature-based solutions: Opportunities and challenges for water treatment, Elsevier eBooks, pp. 575–596, 2025, https://doi.org/10.1016/b978-0-443-21794-4.00011-9.
7. Rajput, V.D., Kumari, S., Minkina, T., Sushkova, S., Mandzhieva, S., Nano-Enhanced Microbial Remediation of PAHs contaminated soil. *Air Soil Water Res.*, 16, 2023, https://doi.org/10.1177/11786221231170099.
8. Arora, D., Arora, A., Panghal, V., Singh, A., Bala, R., Kumari, S., Kumar, S., Unleashing the Feasibility of Nanotechnology in Phytoremediation of Heavy Metal–Contaminated Soil: A Critical Review towards Sustainable

Approach. *Water Air Soil Pollut.*, 235, 1, 2024, https://doi.org/10.1007/s11270-023-06874-9.

9. Rani, P., Rose, P.K., Kidwai, M.K., Meenakshi, Brassica Juncea L.: A Potential Crop for Phytoremediation of Various Heavy Metals, in: *Heavy Metal Toxicity: Environmental Concerns, Remediation and Opportunities*, pp. 285–311, Springer, Singapore, 2023, https://doi.org/10.1007/978-981-99-0397-9_14.
10. Karnwal, A., Martolia, S., Dohroo, A., Al-Tawaha, A.R.M.S., Malik, T., Exploring bioremediation strategies for heavy metals and POPs pollution: the role of microbes, plants, and nanotechnology. *Front. Environ. Sci.*, 12, 2024, https://doi.org/10.3389/fenvs.2024.1397850.
11. Prusty, P., Mohakud, S.K., Mandal, A.R., Parray, K.A., Farooq, S.H., Assessment of heavy metal contaminations in groundwater and surface water resources of Sukinda Valley, Odisha, India. *Geol. J.*, 2025, https://doi.org/10.1002/gj.5227.
12. Kumar, V., Umesh, M., Shanmugam, M.K., Chakraborty, P., Duhan, L., Gummadi, S.N., Pasrija, R., Jayaraj, I., Huligowda, L.K.D., A retrospection on mercury contamination, bioaccumulation, and toxicity in diverse environments: current insights and future prospects. *Sustainability*, 15, 18, 13292, 2023, https://doi.org/10.3390/su151813292.
13. Eltarahony, M., Ibrahim, E., Hegazy, G., Sabry, A., Microbial Remediation of Mercury: An Overview, in: *Environmental Science and Engineering*, pp. 201–234, Springer, Singapore, 2023, https://doi.org/10.1007/978-981-99-7719-2_8.
14. Nain, R., Patel, H., Chahar, M., Kumar, S., Rohilla, D., Pal, M., Biosynthesized metallic nanoparticles for sustainable environmental remediation: mechanisms, applications, and future perspectives. *Discov. Chem.*, 2, 1, 2025, https://doi.org/10.1007/s44371-025-00203-1.
15. Verma, S., Bhatt, P., Verma, A., Mudila, H., Prasher, P., Rene, E.R., Microbial technologies for heavy metal remediation: effect of process conditions and current practices. *Clean Technol. Environ. Policy*, 25, 5, 1485–1507, 2023, https://doi.org/10.1007/s10098-021-02029-8.
16. Pattanayak, D., Pal, D., Thakur, C., Kumar, S., Devnani, G., Bio-synthesis of iron nanoparticles for environmental remediation: Status till date. *Mater. Today Proc.*, 44, 3150–3155, 2021, https://doi.org/10.1016/j.matpr.2021.02.821.
17. Dhokpande, S.R., Deshmukh, S.M., Khandekar, A., Sankhe, A., A review outlook on methods for removal of heavy metal ions from wastewater. *Sep. Purif. Technol.*, 350, 127868, 2024, https://doi.org/10.1016/j.seppur.2024.127868.
18. Singh, V., Pant, N., Sharma, R.K., Padalia, D., Rawat, P.S., Goswami, R., Singh, P., Kumar, A., Bhandari, P., Tabish, A., Deifalla, A.M., Adsorption Studies of Pb(II) and Cd(II) Heavy Metal Ions from Aqueous Solutions Using a Magnetic Biochar Composite Material. *Separations*, 10, 7, 389, 2023, https://doi.org/10.3390/separations10070389.
19. Viswanathan, S.P., Kuriakose, G.M., Neelamury, S.P., Njazhakunnathu, G.V., Paili, T.A., Biochar-based composites as environmentally sustainable

functional materials for wastewater treatment, Elsevier eBooks, pp. 303–318, 2023, https://doi.org/10.1016/b978-0-323-96020-5.00199-0.

20. Kumar, A., Dabodiya, T.S., Ramamoorthy, D., Biochar-Based nano-composites for the removal of organic environmental contaminants, in: *Advances in Science, Technology & Innovation/Advances in Science, Technology & Innovation*, pp. 85–92, Springer, Cham, 2023, https://doi.org/10.1007/978-3-031-28873-9_7.
21. Singhal, A., Parihar, A., Khare, V., Tripathi, G.K., Khan, R., Chemical Approaches to Remediate Heavy Metals, in: *Remediation of Heavy Metals*, pp. 123–147, Wiley, 2023. https://doi.org/10.1002/9781119853589.ch7.

Index

Also of Interest

Check out these related titles from Scrivener Publishing

Catalysis from Waste: Transforming Solid Residue into Value, Edited by Deepti Goyal, Sakshi Kabra Malpani, and Renuka Jain, ISBN: 9781394289301. Transforming environmental liabilities into economic assets, this groundbreaking book provides researchers and policymakers with the definitive, research-backed blueprint for using advanced catalysis to turn solid waste streams into high-value resources.

Solid Waste Management: Challenges, Sustainability and Advancements, Edited by Priyanka Singh, Pooja Agarwal, and V. Vivekanand, ISBN: 9781394267934. Tackle the global challenge of environmental sustainability with this essential book, which provides a critical review of innovative and sustainable solid waste management strategies by leveraging advanced technologies like AI and geographic information system-based predictive models.

Waste Management and Utilization for a Sustainable Environment, Edited by Amit Srivastava and Suman Lata Tripathi, ISBN: 9781394292974. *Waste Management and Utilization for a Sustainable Environment* is an essential guide to understanding how cutting-edge technologies and innovative engineering solutions can transform waste management into a cornerstone of a sustainable future.

Sustainable Production Innovations: Bioremediation and Other Biotechnologies, Edited by Alok Patel and Amit Kumar Sharma, ISBN: 9781119791904. Presenting the latest technologies and practices in this ever-changing field, this groundbreaking new volume covers the gambit for providing solutions and practical applications of smart and efficient energy systems.

Ecosystem Management: Climate Change and Sustainability, Edited by Arnab Banerjee, Manoj Kumar Jhariya, Abhishek Raj, and Taher Mechergui, ISBN: 9781394231218. This book is essential for anyone who wants to understand the challenges of environmental degradation and learn about the sustainable solutions needed to address these critical issues.

Sewage and Biomass from Wastewater to Energy: Possibilities and Technology, Edited by Inamuddin, Tariq Altalhi, Mohammad Luqman, and Joseph K. Bwapwa, ISBN: 9781394204311. Written and edited by a team of industry experts, this exciting new volume covers clean energy production from sewage and biomass while achieving a zero-waste strategy.

Biotechnological Processes for Industrial and Biomedical Waste Management, Edited by Naveen Dwivedi, Shubha Dwivedi, Ranjeet Kumar Mishra and Charles Oluwaseun Adetunji, ISBN: 9781394272006.

LAND AND ENVIRONMENTAL MANAGEMENT THROUGH FORESTRY, Edited by Abhishek Raj, Manoj Kumar Jhariya, Arnab Banerjee, Sharad Nema, and Kiran Bargali, ISBN 9781119910404. Written and edited by a group of experts in the field, this groundbreaking reference work sets the standard for engineers, students, and professionals working in forestry, agriculture, ecology and environmental science, offering the scientific community a way towards combating climate change.

GREEN CHEMISTRY FOR ENVIRONMENTAL REMEDIATION, Edited by Rashmi Sanghi and Vandana Singh, ISBN: 9780470943083. Presents high quality research papers as well as in depth review articles on the new emerging green face of multidimensional environmental chemistry.

BIOREMEDIATION AND SUSTAINABILITY: Research and Applications, Edited by Romeela Mohee and Ackmez Mudhoo, ISBN: 9781118062845. Bioremediation and Sustainability is an up-to-date and comprehensive treatment of research and applications for some of the most important low-cost, "green," emerging technologies in chemical and environmental engineering.

EMERGENCY RESPONSE MANAGEMENT OF OFFSHORE OIL SPILLS: Guidelines for Emergency Responders, By Nicholas P. Cheremisinoff, PhD, and Anton Davletshin, ISBN: 9780470927120. The first book to examine the Deepwater Horizon disaster and offer processes for safety and environmental protection.

COMPLEX BIOLOGICAL SYSTEMS: Adaptation and Tolerance to Extreme Environments, By Irina R. Fomina, Karl Y. Biel, and Vladislav G. Soukhovolsky, ISBN: 9781119510406. Written and edited by some of the most well-respected authors in the area of the adaptation of plants and animals to climate change, this groundbreaking new work is an extremely important scientific contribution to the study of global warming.

GLOBAL CLIMATE CHANGE DEMYSTIFIED, By James G. Speight, ISBN: 9781119653851. Tackling one of the most controversial subjects of our time, one of the world's foremost environmental and petroleum engineers explores the potential causes and ramifications of global climate change.

DEATH OF THE CHESAPEAKE: A History of the Military Role in Polluting the Bay, By Richard D. Albright, ISBN: 9781118686270. This unique book focuses attention on the failure of current efforts to cleanup the Chesapeake Bay and suggests an approach often used in cleaning up environmentally damaged sites

Printed and bound by CPI Group (UK) Ltd, Croydon, CR0 4YY

01/10/2026

14973771-0002